PLANT STRESS PHYSIOLOGY

Edited by
Cajumban, Alyssa A.

ArclerPress

Plant Stress Physiology

Edited by Cajumban, Alyssa A.

Edition 2018

ArclerPress

Published by
Arcler Press LLC
708 3rd Avenue, 6th Floor
New York
NY 10017
United States of America
www.arclerpress.com

ISBN: 978-1-68094-534-8 (Hardback)

Printed in United States of America

Library of Congress Control Number: 2017933106

About the Editor

Cajumban, Alyssa A.

Alyssa Cajumban graduated with a degree in Biology at the University of the Philippines - Manila. In the broad field of biological sciences, her interests focus on microbiology and plant pathology. Some of her works include researching and writing scientific papers regarding fungi, diseases of agricultural crops, and post-harvest techniques. She also works as an editorial assistant and writer for a scientific journal and textbook publication

Table of Contents

3 HEAT STRESS TOLERANCE

4 HARDINESS OF PLANTS

7 FROST AND HEAVY METAL TOLERANCE IN PLANTS

List of Abbreviations

OA	Osmotic Adjustment
ODT/ha	Oven-Dry Tonnes Per Hectare
SMEC	Show-Me-Energy Cooperative
ORAC	Oxygen Radical Absorbance Capacity
P	Phosphorus
PSII	Photosystem II
ROS	Reactive Oxygen Species
RWC	Relative Water Content
UV	Ultraviolet Light
USDA	United States Department of Agriculture
SOD	Superoxide Dismutase
ARS	Agricultural Research Service
AHS	American Horticultural Society
ABA	Abscisic Acid
CTD	Canopy Temperature Depression
CMS	Cell Membrane Stability
Tc	Critical Temperature
GBHs	Glyphosate-Based Herbicides
HSPs	Heat Shock Proteins
INTA	Instituto Nacional de Tecnología Agropecuaria
IPCC	Intergovernmental Panel on Climatic Change
LEDs	Light-Emitting Diodes
NHEJ	non-Homologous Enjoining
NO	Nitric Oxide

Preface

Energy is an absolute requirement for the maintenance of structural organization during the lifetime of an organism. The maintenance of such complex order over time requires a constant input of energy. This results to a constant flow of energy through all biological organisms, which provides the dynamic driving force for the performance of important maintenance processes such as cellular biosynthesis and transport to maintain its characteristic structure and organization as well as the capacity to replicate and grow.

The maintenance of a steady-state results to homeostasis, the meta-stable condition of organisms. Any change in the surrounding environment of an organsim may disrupt its homeostasis.When environmental factors cause the modulation of homeostasis, the response may be defined as biological stress.Thus, it follows that plant stress can be defined as the adverse effect on the physiology of a plant, which is induced upon a sudden transition from some optimal environmental condition, where homeostasis is maintained, to a suboptimal condition, which disrupts the initial homeostatic state.

Plant stress is considered a relative term because the experimental design to assess the impact of a stress always involves the measurement of a physiological phenomenon in a plant species under a suboptimal, stress condition compared to the measurement of the same physiological phenomenon in the same plant species under optimal conditions. Plant stress responses describe the suite of molecular and cellular processes that are triggered when a plant detects some form of stress. Stresses can be abiotic, such as drought or excess light, or biotic, such as herbivores or pathogens. The current concept of stress in plants has been well developed over the past 60 years. Any unfavorable condition or substance that affects or blocks a plant's metabolism, growth, or development is regarded as stress. Vegetation stress can be induced by various natural and anthropogenic stress factors. It must be noted that one has to differentiate between short-term and long-term stress effects as well as between low-stress events that can be partially compensated for by acclimation, adaptation, and repair mechanisms, on the one hand, and strong stress or chronic stress events causing considerable damage that may eventually lead to cell and plant death. In this textbook, some essential stress syndrome responses of plants are summarized in a unifying stress concept. The major abiotic, biotic, and anthropogenic stressors are explained. Some stress tolerance mechanisms are mentioned.

Stress conditions and stress-induced damage in plants have so far been detected using the classical ecophysiological field methods as well as point data measurements of particular chlorophyll fluorescence parameters and of reflectance spectra. The novel laser-induced high-resolution fluorescence imaging technique, which integrates chlorophyll and blue-green fluorescence, marks a new standard in the detection of stress in plants.

Content Coverage

Chapters One and *Two* present the drought tolerance behavior and biotic & abiotic stress factors, correspondingly. Drought tolerance refers to the degree to which a plant is adapted to arid or drought conditions. Biotic stress factors include insect pests and diseases, whereas plant abiotic stress results from environmental factors.

Chapters Three and *Four* present a coverage of heat stress tolerance and hardiness of plants, consistently. Heat stress and the tolerance mechanisms are characterized along with the main characteristics associated with heat stress, On the other hand, hardiness of plants describes their ability to survive adverse growing conditions.

Chapters Five and *Six* focus on desiccation tolerance and salt tolerance of plants, separately. Desiccation is the state of extreme dryness, or the process of extreme drying. Meanwhile, a halophyte is a plant that grows in waters of high salinity, coming into contact with saline water through its roots or by salt spray, such as in saline semi-deserts, mangrove swamps, marshes and sloughs and seashores.

Chapters Seven and *Eight* elucidate the details of frost and soil salinity tolerance in plants, separately. Frost is known to damage crops or reduce future crop yields, therefore, farmers in regions where frost is a problem often invest substantial means to prevent its formation. Salts occur naturally within soils and water. Salination can be caused by natural processes, such as mineral weathering, or by the gradual withdrawal of an ocean. It can also come about through artificial processes, such as irrigation.

CHAPTER 1

Drought Tolerance

Chapter Contents

Drought tolerance is a learned behavior. It is understood by mature native plants. On the other hand, young plants, such as those in plant nurseries, still cannot handle dry summers by themselves. These young plants need about two to three years before they can get accustomed to their environment and survive the summer season without any assistance, such as artificial watering done by humans. However, during extremely hot and dry consecutive years, even adult native plants need help because, all plants do need water, to survive. Nonetheless, during the wet seasons, plants can naturally obtain water from the environment and thus, there is usually no need for artificial watering.

INTRODUCTION

Drought tolerance refers to the degree to which a plant is adapted to arid conditions. Desiccation tolerance, on the other hand, is an extreme degree of drought tolerance. Plants naturally adapted to dry conditions are called xerophytes.

Sedum is a drought tolerant plant whose specific adaptations include succulence and a waxy surface on its leaves and stems.

XEROPHYTE

A xerophyte (from Greek xero dry, phuton plant) is a species of plant that has adapted to survive in an environment with little liquid water, such as a desert or an ice- or snow-covered region in the Alps or the Arctic.

The morphology and physiology of xerophytes are variously adapted to conserve water, and commonly also to store large quantities of water, during dry periods. Other species may be adapted to survive long periods of desiccation of their tissues, during which their metabolic activity may effectively shut down. Plants with such morphological and physiological adaptations are xeromorphic.

The Joshua tree is an example of a xerophyte.

Plants absorb water from the soil, which then evaporates from their outer surfaces; this process is known as transpiration.In dry environments, the rate of evaporation of water from a typical (mesophytic) plant is faster than the rate of absorbtion of water from the soil, ultimately leading to wilting of the plant. To reduce this effect, xerophytic plants exhibit a variety of specialized adaptations to survive in such conditions. Xerophytes may use water from their own storage, allocate water specifically to sites of new tissue growth, or lose less water to the atmosphere and so convert a greater proportion of water in the soil to growth, or have other adaptations to manage water supply and enable them to survive.

Cacti and other succulents are commonly found in deserts, where there is little rainfall. Other xerophytes, such as certain bromeliads, can survive through both extremely wet and extremely dry periods and can be found in seasonally moist habitats such as tropical forests, exploiting niches where water supplies are too intermittent for mesophytic plants to survive. Likewise, chaparral plants are adapted to Mediterranean climates, which have wet winters and dry summers. Plants that live under arctic conditions also have a need for xerophytic adaptations, since water is unavailable for uptake when the ground is frozen.

Types of Xerophytes

Succulent plants store water in their stems or leaves. They include the Cactaceae family, which has round stems and can store a lot of water. The leaves are often vestigial, as in the case of cacti, wherein the leaves are reduced to spines.

Water is stored in the bulbs of some plants, at or below ground level. They may be dormant during drought conditions and are, therefore, known as drought evaders.

Importance of Water Conservation

If the water potential (or strictly, water vapour potential) inside a leaf is higher than outside, the water vapour will diffuse out of the leaf down this gradient. This loss of water vapour from the leaves is called transpiration. Transpiration is natural and inevitable for plants, and much water is lost through this process. Moreover, water is important to plants because it keeps plant cells turgid and alive. When a plant loses too much water, its cells will lose turgor and undergo plasmolysis. This will ultimately lead to wilting, and eventually, death of the plant. Therefore, it is important that plants, especially those that live in dry conditions, have adaptations that can reduce water loss and decrease the size of the open stomata.

In brief, the rate of transpiration is governed by the number of stomata, leaf area (allowing for more stomata), temperature differential, the relative humidity, the presence of wind or air movement, the light intensity, and the presence of a waxy cuticle. It is important to note, that whilst it is vital to keep stomata closed, they have to be opened for gaseous exchange in photosynthesis.

Morphological Adaptations

Cereus peruvianus

Euphorbia virosa

The cactus *Cereus peruvianus* looks superficially very similar to *Euphorbia virosa* due to convergent evolution.

Xerophytic plants may have similar shapes, forms, and structures and look very similar, even if the plants are not very closely related, through a process called convergent evolution. For example, some species of cacti (members of the family *Cactaceae*), which evolved only in the Americas, may appear similar to *Euphorbia*s, which are distributed worldwide. An unrelated species of caudiciforms, plants with swollen bases that are used to store water, may also display such similarities.

Reduction of Surface Area

Reducing the plant's surface area exposed to air can reduce water loss caused by evaporation. Thus, xerophytes can have smaller leaves, fewer branches, and overall less surface area than other plants. An example of leaf surface reduction are the spines of a cactus. An example of compaction and reduction of branching are the barrel cacti. Other xerophytes may have their leaves compacted at the base, as in a basal rosette, which may be smaller than the plant's flower. This adaptation is exhibited by some *Agave* and *Eriogonum* species, which can be found growing near Death Valley.

Reduction in Air Flow

Some xerophytes have tiny hairs on their surface to provide a wind break and reduce air flow, thereby reducing the rate of evaporation. When a plant surface is covered with tiny hairs, it is called tomentose.

In a still environment, the areas under the leaves/spines where transpiration is taking place form a small localised environment that is more saturated than normal with water vapour. If this is not blown away by wind, the water vapour potential gradient is reduced and so is transpiration. Thus, in a windier situation, this localization is not held and so the gradient remains high, which aids the loss of water vapour. Spines trap a layer of moisture and also slow air movement over tissues.

Reflectivity

The color of a plant, or of the waxes or hairs on its surface, may serve to reflect sunlight and reduce evaporation. An example is the white chalky wax (epicuticular wax) coating of *Dudleya brittonii*, which has the highest ultraviolet light (UV) reflectivity of any known naturally occurring biological substance.

Physiological adaptations

Some plants can store water in root structures, trunk structures, stems, and leaves. Water storage in swollen parts of the plant is known as succulence. A swollen trunk or root at the ground level of a plant is called a caudex and plants with swollen bases are called caudiciforms.

Tiny pores on the surface of a xerophytic plant called stomata may open only at night, so as to reduce evaporation.

Plants may secrete resins and waxes (epicuticular wax) on their surfaces, which reduce evaporation. Examples are the heavily scented and flammable resins (volatile organic compounds) of some chaparral plants, such as *Malosma laurina*, or the chalky wax of *Dudleya pulverulenta*.

Plants may drop their leaves in times of dryness (drought deciduous), or modify the leaves produced so that they are smaller.

During dry times, xerophytic plants may stop growing and go dormant, or change the allocation of the products of photosynthesis from growing new leaves to the roots.

Seeds may be modified to require an excessive amount of water before germinating, so as to ensure a sufficient water supply for the seedling's survival. An example of this is the California poppy, whose seeds lie dormant during drought and then germinate, grow, flower, and form seeds within four weeks of rainfall.

Modification of Environment

The leaf litter on the ground around a plant can provide an evaporative barrier to prevent water loss. A plant's root mass itself may also hold organic material that retains water, as in the case of the arrowweed (*Pluchea sericea*).

Mechanism Table

Mechanism	Adaptation	Example
Limit water loss	waxy stomata	prickly pear
	few stomata	
	sunken stomata	pine
	stomata open at night	tea plant
	CAM photosynthesis	cactus
	large hairs on surface	Bromeliads
	curled leaves	esparto grass
Storage of water	succulent leaves	*Kalanchoe*
	succulent stems	*Euphorbia*
	fleshy tuber	*Raphionacme*
Water uptake	deep root system	*Acacia, Prosopis*
	below water table	*Nerium oleander*
	absorbing surface moisture from leaf hairs or trichomes	*Tillandsia*

BREEDING FOR DROUGHT STRESS TOLERANCE

Breeding for drought stress tolerance is the process of breeding plants with the goal of reducing the impact of drought on plant growth.

Drought Stress in Crop Plants

In nature or crop fields, water is often the most limiting factor for plant growth. If plants do not receive adequate rainfall or irrigation, the resulting drought stress can reduce growth more than all other environmental stresses combined.

Drought can be defined as the absence of rainfall or irrigation for a period of time sufficient to deplete soil moisture and injure plants. Drought stress results when water loss from the plant exceeds the ability of the plant's roots to absorb water and when the plant's water content is reduced enough to interfere with normal plant processes.

Drought Stress is a Global Phenomenon

About 15 million km^2 of the land surface is covered by cropland, and about 16% of this area is equipped for irrigation (Siebert *et al.* 2005). Thus in many parts of the world including the United States, plants may frequently encounter drought stress. Rainfall is very seasonal and periodic drought occurs regularly. The effect of drought is more prominent in sandy soils with low water holding capacity. On such soils some plants may experience drought stress after only a few days without water.

During the last century, the rate of increase in `blue' water withdrawal (from rivers, lakes, and aquifers) for irrigation and other purposes was higher than the growth rate of the world population (Shiklomanov 1998). Country-wise maps of irrigated areas are available.

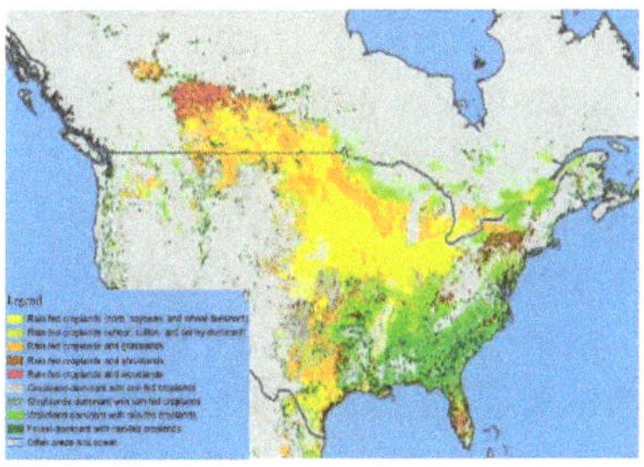

Rain fed areas of USA, for details - Biradar et al.(2009) International J Applied Earth Observation and Geoinformation 11: 114-129

Drought stress and future challenges to crop production

Moisture deficits is a significant challenge to the future crop production. Severe drought in parts of the U.S., Australia, and Africa in recent years drastically reduced crop yields and disrupted regional economies. Even in average years, however, many agricultural regions, including the U.S. Great Plains, suffer from chronic moisture deficits. Cereal crops typically attain only about 25% of their potential yield due to the effects of environmental stress, with moisture stress the most important cause. Two major trends will likely increase the frequency and severity of crop moisture deficits:

Global climate change.

Higher temperatures are likely to increase crop water use due to increased transpiration. A warmer atmosphere will also speed up melting of mountain snowpack, resulting in less water available for irrigation. More extreme weather patterns will increase the frequency of drought in some regions.

Competition in using limited water supplies.

Increased demand from municipal and industrial users will further reduce the amount of water available for irrigated crops.

Although changes in tillage and irrigation practices can improve production by conserving water, enhancing the genetic tolerance of crops to drought stress is considered an essential strategy for addressing moisture deficits.

Drought Stress Affects Plant Physiology

A plant responds to a lack of water by halting growth and reducing photosynthesis and other plant processes in order to reduce water use. As water loss progresses, leaves of some species may appear to change color—usually to blue-green. Foliage begins to wilt and, if the plant is not irrigated, leaves will fall off and the plant will eventually die. Drought lowers the water potential of a plant's root and upon extended exposure, abscisic acid is accumulated and eventually stomatal closure occurs. This reduces a plant's leaf relative water content. The time required for drought stress to occur depends on the water-holding capacity of the soil, environmental conditions, stage of plant growth, and plant species. Plants growing in sandy soils with low water-holding capacity are more susceptible to drought stress than plants growing in clay soils. A limited root system will accelerate the rate at which drought stress develops. A root system may be limited by the presence of competing root systems, by site conditions such as compacted soils or high water tables, or by container size (if growing in a container). A plant with a large mass of leaves in relation to the root system is prone to drought stress because the leaves may lose water faster than the roots can supply it. Newly installed plants and poorly established plants may be especially susceptible to drought stress because of the limited root system or the large mass of stems and leaves in comparison to roots.

Drought Stress Interaction with other Stress Factors

Aside from the moisture content of the soil, environmental conditions of high light intensity, high temperature, low relative humidity and high wind speed will significantly increase plant water loss. The prior environment of a plant also can influence the development of drought stress. A plant that has been drought stressed previously and has recovered may become more drought resistant. Also, a plant that was well-watered prior to drought will usually survive drought better than a continuously drought-stressed plant.

Mechanism of Drought Stress Resistance

Degree of Resistance to drought depend upon crops. Generally three strategies can help a crop to mitigate the effect of drought stress:

(a) Drought escape
(b) Drought avoidance
(c) Drought tolerance

A proper timing of lifecycle, resulting in the completion of the most sensitive developmental stages while water is abundant, is considered to be a drought escape strategy. Avoiding water-deficit stress with a root system capable of extracting water from deep soil layers, or by reducing evapotranspiration without affecting yields, is considered as drought avoidance. Mechanisms such as osmotic adjustment (OA) whereby a plant maintains cell turgor pressure under reduced soil water potential are categorized as drought tolerance mechanisms. Drought avoidance mechanisms can be expressed even in the absence of stress and are then considered constitutive. Drought tolerance mechanisms are the result of a response triggered by drought stress itself and are therefore considered adaptative. When the stress is terminal and predictable, drought escape through the use of shorter duration varieties is often the preferable method of improving yield potential. Drought avoidance and tolerance mechanisms are required in situations where the timing of drought is mostly unpredictable.

Drought tolerance mechanism is genetically controlled and genes or QTL responsible for drought tolerance have been discovered in several crops which opens avenue for molecular breeding for drought tolerance.

Drought Tolerance Traits

Tolerance to drought is a quantitative trait, with a complex phenotype, often confounded by plant phenology. Breeding for drought tolerance is further complicated since several types of abiotic stress, such as high temperatures, high irradiance, and nutrient toxicities or deficiencies can challenge crop plants simultaneously.

(a) Osmotic adjustment When a plant is exposed to water deficit, it may accumulate a variety of osmotically active compounds such as amino acids and sugars, resulting in a lowering of the osmotic potential. Examples of amino acids that may be up-regulated are proline and glycine betaine. This is termed osmotic adjustment and enables the plant to take up water, maintain turgor and survive longer.

(b) Cell membrane stability The ability to survive dehydration is influenced by a cell's ability to survive at reduced water content. This can be considered complementary to OA because both traits will help maintain leaf growth (or prevent leaf death) during drought. Crop varieties differ in dehydration tolerance and an important factor for such differences is the capacity of the cell membrane to prevent electrolyte leakage at decreasing water content, or "cell membrane stability (CMS)". The maintenance of membrane function is assumed to mean that cell activity is also maintained. Measurements of CMS have been used in different crops and are known to be correlated with yields under high temperature and possibly under drought stress.

(c) Epicuticular wax In sorghum (Sorghum bicolor L. Moench), drought resistance is a trait that is highly correlated with the thickness of the epicuticular wax layer. Experiments have demonstrated that rice varieties with a thick cuticle layer retain their leaf turgor for longer periods of time after the onset of a water-stress.

(d) Partitioning and stem reserve mobilization

As photosynthesis becomes inhibited by drought, the grain filling process becomes increasingly reliant on stem reserve utilization.72 Numerous studies have reported that stem reserve mobilization capacity is related to yield under water-stress in wheat. In rice, a few studies also indicated that this mechanism maintains grain yield under water stress at the grain filling stage. This drought tolerance mechanism is stimulated by a decrease in gibberellic acid concentration and an increase in abscisic acid concentration.

(e) Manupulation and Stability of flowering processes

Seeds emerge from deep sowing, exploiting the dry upper soil, in order to reach the receding moisture profile and avoid high soil surface temperatures that inhibit germination. Screening at these stage is advantageous, specially when managing large amount of germplasms.

Crop Ideotype for Drought Tolerance

Usually ideotypes are developed to create an ideal plant variety. The following traits constitutes ideotype of wheat by CIMMYT.

1) Large seed size. Helps emergence, early ground cover, and initial biomass.
2) Long coleoptiles. For emergence from deep sowing
3) Early ground cover. Thinner, wider leaves (i.e., with a relatively low specific leaf weight) and a more prostrate growth habit help to increase ground cover, thus conserving soil moisture and potentially increasing radiation use efficiency.
4) High pre-anthesis biomass.
5) Good capacity for stem reserves and remobilization
6) High spike photosynthetic capacity
7) High RLWC/Gs/CTD during grain filling to indicate ability to extract water
8) **Osmotic adjustment**

(9) **Accumulation of ABA:-**The benefit of ABA accumulation under drought has been demonstrated (Innes et al. 1984). It appears to pre-adapt plants to stress by reducing stomatal conductance, rates of cell division, organ size,and increasing development rate. However, high ABA can also result in sterility problems since high ABA levels may abort developing florets

10) **Heat Tolerance.** The contribution of heat tolerance to performance under moisture stress needs to be quantified, but it is relatively easy to screen for (Reynolds et al. 1998).

11) **Leaf anatomy:** waxiness, pubescence, rolling, thickness, posture. These traits decrease radiation load to the leaf surface. Benefits include a lower

evapotranspiration rate and reduced risk of irreversible photo-inhibition. However, they may also be associated with reduce radiation use efficiency, which would reduce yield under more favorable conditions.

12) **High tiller survival:-** Comparison of old and new varieties have shown that under drought older varieties over-produce tillers many of which fail to set grain while modern drought tolerant lines produce fewer tillers most of which survive.

13) **Stay-green:-** The trait may indicate the presence of drought avoidance mechanisms, but probably does not contribute to yield per se if there is no water left in the soil profile by the end of the cycle to support leaf gas exchange. It may be detrimental if it indicates lack of ability to remobilize stem reserves. However, research in sorghum has indicated that staygreen is associated with higher leaf chlorophyll content at all stages of development and both were associated with improved yield and transpiration efficiency under drought.

Combination Phenomics : Overall Health of Crops

The concept of combination phenomics comes from the idea that two or more stresses have common physiological effects or common traits - which are an indicator of overall plant health. Similar analogy in human medical terms is high blood pressure or high body temperature or high white blood cells in body is an indicator of health problems and thus we can select healthy people from unhealthy using such a measure. As both abiotic and abiotic stresses can result in similar physiological consequence, tolerant plant can be separated from sensitive plants. Some imaging or infrared measuring techniques can help to speed the process for breeding process. For example spot blotch intensity and canopy temperature depression can be monitored with canopy temperature depression.

Molecular Breeding for Drought Tolerance

Recent research breakthroughs in biotechnology have revived interest in targeted drought tolerance breeding and use of new genomics tools to enhance crop water productivity. Marker-assisted breeding is revolutionizing the improvement of temperate field crops and will have similar impacts on breeding of tropical crops. Other molecular breeding tool include development of genetically modified crops that can tolerate plant stress. As a complement to the recent rapid progress in genomics, a better understanding of physiological mechanisms of drought response will also contribute to the progress of genetic enhancement of crop drought tolerance. It is now well accepted that the complexity of the drought syndrome can only be tackled with a holistic approach that integrates physiological dissection of crop drought avoidance and tolerance traits using molecular genetic tools such as MAS, microarrays and transgenic crops, with agronomic practices that lead to better conservation and utilization of soil moisture, and better matching of crop genotypes with the environment.

REFERENCES

1. Ramankutty N, Evan A T, Monfreda C and Foley J A 2008 Farming the planet. Part 1: the geographic distribution of global agricultural lands in the year 2000 Glob. Biogeochem. Cycles 22 GB1003
2. Siebert, S; Döll, P; Hoogeveen, J; Faures, J M; Frenken, K; Feick, S (2005). "Development and validation of the global map of irrigation areas". Hydrol. Earth Syst. Sci. 9: 535–47. doi:10.5194/hess-9-535-2005.
3. Shiklomanov I A 1998 World Water Resources—A New Appraisal and Assessment for the 21st Century
4. Stefan Siebert, Petra Döll, Sebastian Feick, Jippe Hoogeveen and Karen Frenken (2007) Global Map of Irrigation Areas version 4.0.1. Johann Wolfgang Goethe University, Frankfurt am Main, Germany / Food and Agriculture Organization of the United Nations, Rome, Italy
5. http://www.sciencedirect.com/cache/MiamiImageURL/1-s2.0-S0303243408000834-gr7_lrg.jpg/0?wchp=dGLzVlk-zSkzk
6. Ogbaga, Chukwuma C.; Stepien, Piotr; Johnson, Giles N. (October 2014). "Sorghum (Sorghum bicolor) varieties adopt strongly contrasting strategies in response to drought". Physiologia Plantarum. 152 (2): 389–401. doi:10.1111/ppl.12196.
7. Rosyara, U.R., A. A. Ghimire, S. Subedi, and Ram C. Sharma. 2008.Variation in south Asian wheat germplasm for seedling drought tolerance traits. Plant Genetic Resources 6: 88-93
8. Richards, R.A. (1996). "Defining selection criteria to improve yield under drought". Plant Growth Regulation. 20: 157–166. doi:10.1007/bf00024012.
9. Innes, P., R.D. Blackwell, and S.A. Quarrie. 1984. Some effects of genetic variation in drought-induced abscisic acid accumulation on the yield and water-use of spring J. Agric. Sci. Camb. 102:341
10. Reynolds, M.P., R.P. Singh, A. Ibrahim, O.A.A. Ageeb, A. Larque-Saavedra, and J.S. Quick. 1998. Evaluating physiological traits to complimentempirical selection for wheat in warm environments. Euphytica 100:85-9
11. Rosyara, U.R., Vromman, D., Duveiller, E. 2008. Canopy temperature depression as indication of correlative measure of spot blotch resistance and heat stress tolerance in spring wheat. J. Plant Path. 90 :103–107.
12. Rosyara, U. R., S. Subdedi, R. C. Sharma and E. Duveiller.2010.The effect of spot blotch and heat stress in variation of canopy temperature depression, chlorophyll fluorescence and chlorophyll content of hexaploid wheat genotypes. Euphytica Volume 174, Number 3, 377-390
13. Rosyara, U. R., S. Subdedi, R. C. Sharma and E. Duveiller. 2010. Photochemical Efficiency and SPAD Value as Indirect Selection Criteria for Combined Selection of Spot Blotch and Terminal Heat Stress in Wheat. Journal of Phytopathology Volume 158, Issue 11-12, pages 813–821
14. Rosyara, U.R., Vromman, D., Duveiller, E. 2008. Canopy temperature depression

as indication of correlative measure of spot blotch resistance and heat stress tolerance in spring wheat. J. Plant Path. 90 :103–107.

15. "Xeromorphic", The Cambridge Illustrated Glossary of Botanical Terms, Michael Hickey, Clive King, Cambridge University Press, 2001
16. Mulroy, Thomas W. (1979). "Spectral properties of heavily glaucous and non-glaucous leaves of a succulent rosette-plant". Oecologia. 38 (3): 349–357. doi:10.1007/BF00345193.
17. "Plant Adaptations". University of New Mexico. Archived from the original on January 4, 2015. Retrieved December 2, 2014.
18. Chown, S.L.,; Nicolson, S.W. (2004). "Insect Physiological Ecology". New York: Oxford University Press.
19. Gray, E.M.; Bradley, T.J. (2005). "Physiology of desiccation resistance in Anopheles gambiae and Anopheles arabiensis". American Journal of Tropical Medicine and Hygiene. 73 (3): 553–559. PMID 16172480.
20. Graves, J. L.; Toolson, E. C.; Jeong, C.; Vu & L.N.; Rose, M.R. (1992). "Desiccation, flight, glycogen and postponed senescence in Drosophila melanogaster". Physiological Zoology. 65: 268–286.
21. Gefen, E.; Marlon, A.J.; Gibbs, A.G. (2006). "Selection for desiccation resistance in adult Drosophila melanogaster affects larval development and metabolite accumulation". Journal of Experimental Biology. 209 (Pt 17): 3293–3300. doi:10.1242/jeb.02397. PMID 16916965.
22. Folk, D.G.; Han, C.; Bradley, T.J. (2001). "Water acquisition and partitioning in Drosophila melanogaster: effects of selection for desiccation-resistance". Journal of Experimental Biology. 204 (Pt 19): 3323–3331. PMID 11606606.
23. Mellanby, K. (1934). "The evaporation of water from insects". Journal of Experimental Biology. 10: 317–333. doi:10.1111/j.1469-185x.1935.tb00487.x.
24. Machin J. 1968. The Permeability of the Epiphragm of Terrestrial Snails to Water Vapor. Biological Bulletin, Vol. 134, No. 1 (Feb., 1968), pp. 87-95
25. Wigglesworth, V.B. (1984). "Insect Physiology". Chapman and Hall, NY.
26. Gibbs, A.G. (2002). "Lipid melting and cuticular permeability: new insights into an old problem". Journal of Insect Physiology. 48 (4): 391–400. doi:10.1016/S0022-1910(02)00059-8. PMID 12770088.
27. Treherne, J.E.; Willmer, P.G. (1975). "Hormonal control of integumentary water loss: evidence for a novel neuroendocrine system in an insect (Periplaneta americana)". Journal of Experimental Biology. 63 (1): 143–159. PMID 1159358.
28. Snyder, G.K.; Sheafor, B.; Scholnick, D.; Farrelly, C. (1995). "Gas Exchange in the Insect Tracheal System". Journal of Theoretical Biology. 172 (3): 199–207. doi:10.1006/jtbi.1995.0016. PMID 7715192.
29. Bursell, E. (1960). "Loss of water by excretion and defaecation in the Tsetse fly". Experimental Biology. 37: 689–697.

30. Willmer, P.; Stone, G.; Johnston, I.A. (2000). "Environmental physiology of animals". New York: Blackwell Publishing.
31. Renault, D.; Coray, Y. (2004). "Water loss of male and female Alphitobius diaperinus (Coleoptera: Tenebrionidae) maintained under dry conditions". European Journal of Entomology. 101: 491–494. doi:10.14411/eje.2004.069.
32. Prince, G.J.; Parsons, P.A. (1977). "Adaptive Behaviour of Drosophila Adults in Relation to Temperature and Humidity". Australian Journal of Zoology. 25 (2): 285–290. doi:10.1071/ZO9770285.
33. Halffter, G.; Edmonds, W.D. (1982). "The Nesting Behavior of Dung Beetles (Scarabaeinae) an Ecological and Evolutive Approach". Mexico D.F.: Instituto de Ecologia.
34. Rowley, M.; Hanson, F. (2007). "Humidity detection and hygropreference behavior in larvae of the tobacco hornworm, Manduca sexta". Journal of Insect Science. 7: 39. doi:10.1673/031.007.3901. PMC 2999434 . PMID 20302460.
35. Watanbe, M.; Kikawada, T.; Okuda, T. (2003). "Increase of internal ion concentration triggers trehalose synthesis associated with cryptobiosis in larvae of Polypedilum vanderplanki". Journal of Experimental Biology. 206 (Pt 13): 2281–2286. doi:10.1242/jeb.00418. PMID 12771176.

CHAPTER 2

Biotic and Abiotic

Chapter Contents

Abiotic stress is a term used to describe nonliving factors that adversely affect living organisms. Animals can be affected by abiotic stress, but plants are more vulnerable as they are unable to move to a less stressful environment. Biotic stress factors would include insect pests and diseases. On the other hand, plant abiotic stress results from environmental factors, which may be related to climate (drought, extremes of temperature and wind, for example) or to chemical factors in the soil and the atmosphere.

INTRODUCTION

Biotic stress is stress that occurs as a result of damage done to plants by other living organisms, such as bacteria, viruses, fungi, parasites, beneficial and harmful insects, weeds, and cultivated or native plants. Not to be confused with abiotic stress, which is the negative impact of non-living factors on the organisms in a specific environment such as sunlight, wind, salinity, over watering and drought. The types of biotic stresses imposed on a plant depend on both geography and climate and on the host plant and its ability to resist particular stresses. Although there are many kinds of biotic stress, the majority of plant diseases are caused by fungi. Biotic stress remains a broadly defined term and those who study it face many challenges, such as the greater difficulty in controlling biotic stresses in an experimental context compared to abiotic stress.

The damage caused by these various living and nonliving agents can appear very similar. Even with close observation, accurate diagnosis can be difficult. For example, browning of leaves on an oak tree caused by drought stress may appear similar to leaf browning caused by oak wilt, a serious vascular disease, or the browning cause by anthracnose, a fairly minor leaf disease.

Agriculture

It is a major focus of agricultural research, due to the vast economic losses caused by biotic stress to cash crops. The relationship between biotic stress and plant yield affects economic decisions as well as practical development. The impact of biotic injury on crop yield impacts population dynamics, plant-stressor coevolution, and ecosystem nutrient cycling.

Biotic stress also impacts horticultural plant health and natural habitats ecology. It also has dramatic changes in the host recipient.Plants are exposed to many stress factors, such as drought, high salinity or pathogens, which reduce the yield of the cultivated plants or affect the quality of the harvested products. *Arabidopsis thaliana* is often used as a model plant to study the responses of plants to different sources of stress.

Biotic stresses have had huge repercussions for humanity; an example of this is the potato blight, an oomycete which caused widespread famine in England, Ireland and Belgium in the 1840s. Another example is grape phylloxera coming from North America in the 19th century, which led to the Great French Wine Blight.

Today

Losses to pests and disease in crop plants continue to pose a significant threat to agriculture and food security. During the latter half of the 20th century, agriculture became increasingly reliant on synthetic chemical pesticides to provide control of pests and diseases, especially within the intensive farming systems common in the developed world. However, in the 21st century, this reliance on chemical control is becoming unsustainable. Pesticides tend to have a limited lifespan due to the emergence of resistance in the target pests, and are increasingly recognised in many cases to have negative impacts on biodiversity, and on the health of agricultural workers and even consumers.

Tomorrow

Due to the implications of climate change, it is suspected that plants will have increased susceptibility to pathogens.Additionally, elevated threat of abiotic stresses (i.e. drought and heat) are likely to contribute to plant pathogen susceptibility.

Effect on Plant Growth

Photosynthesis

Many biotic stresses affect photosynthesis, as chewing insects reduce leaf area and virus infections reduce the rate of photosynthesis per leaf area. Vascular- wilt fungi compromise the water transport and photosynthesis by inducing stomata closure.

Response to Stress

Plants have co-evolved with their parasites for several hundred million years. This co-evolutionary process has resulted in the selection of a wide range of plant defences against microbial pathogens and herbivorous pests which act to minimise frequency and impact of attack. These defences include both physical and chemical adaptations, which may either be expressed constitutively, or in many cases, are activated only in response to attack. For example, utilization of high metal ion concentrations derived from the soil allow plants to reduce the harmful effects of biotic stressors (pathogens, herbivores etc.); meanwhile preventing the infliction of severe metal toxicity by way of safeguarding metal ion distribution throughout the plant with protective physiological pathways. Such induced resistance provides a mechanism whereby the costs of defence are avoided until defense is beneficial to the plant. At the same time, successful pests and pathogens have evolved mechanisms to overcome both constitutive and induced resistance in their particular host species. In order to fully understand and manipulate plant biotic stress resistance, we require a detailed knowledge of these interactions at a wide range of scales, from the molecular to the community level.

Cross Tolerance with Abiotic Stress

- Evidence shows that a plant undergoing multiple stresses, both abiotic and biotic (usually pathogen or herbivore attack), can produce a positive effect on plant performance, by reducing their susceptibility to biotic stress compared to how they respond to individual stresses. The interaction leads to a crosstalk between their respective hormone signalling pathways which will either induce or antagonize another restructuring genes machinery to increase tolerance of defense reactions.
- Reactive oxygen species (ROS) are key signalling molecules produced in response to biotic and abiotic stress cross tolerance. ROS are produced in response to biotic stresses during the oxidative burst.
- Dual stress imposed by ozone (O3) and pathogen affects tolerance of crop and leads to altered host pathogen interaction (Fuhrer, 2003). Alteration in pathogenesis potential of pest due to O3 exposure is of ecological and economical importance.

- Tolerance to both biotic and abiotic stresses has been achieved. In maize, breeding programmes have led to plants which are tolerant to drought and have additional resistance to the parasitic weed *Striga hermonthica*.

Remote Sensing

The Agricultural Research Service (ARS) and various government agencies and private institutions have provided a great deal of fundamental information relating spectral reflectance and thermal emittance properties of soils and crops to their agronomic and biophysical characteristics. This knowledge has facilitated the development and use of various remote sensing methods for non-destructive monitoring of plant growth and development and for the detection of many environmental stresses that limit plant productivity. Coupled with rapid advances in computing and position locating technologies, remote sensing from ground-, air-, and space-based platforms is now capable of providing detailed spatial and temporal information on plant response to their local environment that is needed for site specific agricultural management approaches. This is very important in today's society because with increasing pressure on global food productivity due to population increase, result in a demand for stress-tolerant crop varieties that has never been greater.

BIOTIC COMPONENT

Biotic components are the living things that shape an ecosystem.

Biotic components usually include:

- Producers, i.e. autotrophs: e.g. plants, they convert the energy [from photosynthesis (the transfer of sunlight, water, and carbon dioxide into energy), or other sources such as hydrothermal vents] into food.
- Consumers, i.e. heterotrophs: e.g. animals, they depend upon producers (occasionally other consumers) for food.
- Decomposers, i.e. detritivores: e.g. fungi and bacteria, they break down chemicals from producers and consumers (usually dead) into simpler form which can be reused.

A biotic factor is any living component that affects the population of another organism, or the environment. This includes animals that consume the organism, and the living food that the organism consumes. Biotic factors also include human influence, pathogens and disease outbreaks. Each biotic factor needs energy to do work and food for proper growth.

All species are influenced by biotic factors in one way or another. For example, If the number of predators will increase, the whole food web will be affected (the population number of organisms that are lower in the food web will decrease). Similarly, when organisms have more food to eat, they will grow quicker and will be more likely to reproduce, so the population size will obviously increase. Pathogens and disease outbreaks, however, are most likely to cause a decrease in population size. Humans make the most sudden changes in an environment (e.g. building cities and factories, disposing of waste into the water). These changes are most likely to cause a decrease in the population of any species, due to the sudden appearance of pollutants.

Biotic components are contrasted to abiotic components, which are non-living components that influence population size and the environment. Examples of abiotic factors are: temperature, light intensity, moisture and water levels, air currents, carbon dioxide levels and the pH of water and soil.

The factors mentioned above may either cause an increase or a decrease in population size, depending on the organism. For example, rainfall may encourage the growth of new plants, but too much of it may cause flooding, which may drastically decrease the population size.

ABIOTIC STRESS

Abiotic stress is defined as the negative impact of non-living factors on the living organisms in a specific environment. The non-living variable must influence the environment beyond its normal range of variation to adversely affect the population performance or individual physiology of the organism in a significant way.

Whereas a biotic stress would include such living disturbances as fungi or harmful insects, abiotic stress factors, or stressors, are naturally occurring, often intangible, factors such as intense sunlight or wind that may cause harm to the plants and animals in the area affected. Abiotic stress is essentially unavoidable. Abiotic stress affects animals, but plants are especially dependent on environmental factors, so it is particularly constraining. Abiotic stress is the most harmful factor concerning the growth and productivity of crops worldwide. Research has also shown that abiotic stressors are at their most harmful when they occur together, in combinations of abiotic stress factors.

Examples

Abiotic stress comes in many forms. The most common of the stressors are the easiest for people to identify, but there are many other, less recognizable abiotic stress factors which affect environments constantly.

The most basic stressors include:

- High winds
- Extreme temperatures
- Drought
- Flood
- Other natural disasters, such as tornadoes and wildfires.

Lesser-known stressors generally occur on a smaller scale. They include: poor edaphic conditions like rock content and pH levels, high radiation, compaction, contamination, and other, highly specific conditions like rapid rehydration during seed germination.

Effects

Abiotic stress, as a natural part of every ecosystem, will affect organisms in a variety of ways. Although these effects may be either beneficial or detrimental, the location of the area is crucial in determining the extent of the impact that abiotic stress will have. The higher the latitude of the area affected, the greater the impact of abiotic stress will be on that area. So, a taiga or boreal forest is at the mercy of whatever abiotic stress factors may come along, while tropical zones are much less susceptible to such stressors.

Benefits

One example of a situation where abiotic stress plays a constructive role in an ecosystem is in natural wildfires. While they can be a human safety hazard, it is productive for these ecosystems to burn out every once in a while so that new organisms can begin to grow and thrive. Even though it is healthy for an ecosystem, a wildfire can still be considered an abiotic stressor, because it puts an obvious stress on individual organisms within the area. Every tree that is scorched and each bird nest that is devoured is a sign of the abiotic stress. On the larger scale, though, natural wildfires are positive manifestations of abiotic stress.

What also needs to be taken into account when looking for benefits of abiotic stress, is that one phenomenon may not affect an entire ecosystem in the same way. While a flood will kill most plants living low on the ground in a certain area, if there is rice there, it will thrive in the wet conditions. Another example of this is in phytoplankton and zooplankton. The same types of conditions are usually considered stressful for these two types of organisms. They act very similarly when exposed to ultraviolet light and most toxins, but at elevated temperatures the phytoplankton reacts negatively, while the thermophilic zooplankton reacts positively to the increase in temperature. The two may be living in the same environment, but an increase in temperature of the area would prove stressful only for one of the organisms.

Lastly, abiotic stress has enabled species to grow, develop, and evolve, furthering natural selection as it picks out the weakest of a group of organisms. Both plants and animals have evolved mechanisms allowing them to survive extremes.

Detriments

The most obvious detriment concerning abiotic stress involves farming. It has been claimed by one study that abiotic stress causes the most crop loss of any other factor and that most major crops are reduced in their yield by more than 50% from their potential yield.

Because abiotic stress is widely considered a detrimental effect, the research on this branch of the issue is extensive. For more information on the harmful effects of abiotic stress, see the sections below on plants and animals.

In Plants

A plant's first line of defense against abiotic stress is in its roots. If the soil holding the plant is healthy and biologically diverse, the plant will have a higher chance of surviving stressful conditions.

Facilitation, or the positive interactions between different species of plants, is an intricate web of association in a natural environment. It is how plants work together. In areas of high stress, the level of facilitation is especially high as well. This could possibly be because the plants need a stronger network to survive in a harsher environment, so their interactions between species, such as cross-pollination or mutualistic actions, become more common to cope with the severity of their habitat.

Plants also adapt very differently from one another, even from a plant living in the same area. When a group of different plant species was prompted by a variety of different stress signals, such as drought or cold, each plant responded uniquely. Hardly any of the responses were similar, even though the plants had become accustomed to exactly the same home environment.

Rice (*Oryza sativa*) is a classic example. Rice is a staple food throughout the world, especially in China and India. Rice plants experience different types of abiotic stresses, like drought and high salinity. These stress conditions have a negative impact on rice production. Genetic diversity has been studied among several rice varieties with different genotypes using molecular markers.

Serpentine soils (media with low concentrations of nutrients and high concentrations of heavy metals) can be a source of abiotic stress. Initially, the absorption of toxic metal ions is limited by cell membrane exclusion. Ions that are absorbed into tissues are sequestered in cell vacuoles. This sequestration mechanism is facilitated by proteins on the vacuole membrane.

Chemical priming has been proposed to increase tolerance to abiotic stresses in crop plants. In this method, which is analogous to vaccination, stress-inducing chemical agents are introduced to the plant in brief doses so that the plant begins preparing defense mechanisms. Thus, when the abiotic stress occurs, the plant has already prepared defense mechanisms that can be activated faster and increase tolerance.

Phosphate Starvation in Plants

Phosphorus (P) is an essential macronutrient required for plant growth and development, but most of the world's soil is limited in this important plant nutrient. Plants can utilize P mainly in the form if soluble inorganic phosphate (Pi) but are subjected to abiotic stress of P-limitation when there is not sufficient soluble PO_4 available in the soil. Phosphorus forms insoluble complexes with Ca and Mg in basic soils and Al and Fe in acidic soils that makes it unavailable for plant roots. When there is limited bioavailable P in the soil, plants show extensive abiotic stress phenotype This includes short primary roots, and more lateral roots and root hairs to make more surface area available for PO_4 absorption. Some also exudes organic acids and phosphatase to release the PO_4 complex from P-containing molecules and make it available for growing plants organs. It has been shown that PHR1, a MYB - related transcription factor is a master regulator of P-starvation response in plants. PHR1 also has been shown to regulate extensive remodeling of lipids and metabolites during phosphorus limitation stress

In Animals

For animals, the most stressful of all the abiotic stressors is heat. This is because many species are unable to regulate their internal body temperature. Even in the species that are able to regulate their own temperature, it is not always a completely accurate system. Temperature determines metabolic rates, heart rates, and other very important factors within the bodies of animals, so an extreme temperature change can easily distress the animal's body. Animals can respond to extreme heat, for example, through natural heat acclimation or by burrowing into the ground to find a cooler space.

It is also possible to see in animals that a high genetic diversity is beneficial in providing resiliency against harsh abiotic stressors. This acts as a sort of stock room when a species is plagued by the perils of natural selection. A variety of galling insects are among the most specialized and diverse herbivores on the planet, and their extensive protections against abiotic stress factors have helped the insect in gaining that position of honor.

In Endangered Species

Biodiversity is determined by many things, and one of them is abiotic stress. If an environment is highly stressful, biodiversity tends to be low. If abiotic stress does not have a strong presence in an area, the biodiversity will be much higher.

This idea leads into the understanding of how abiotic stress and endangered species are related. It has been observed through a variety of environments that as the level of abiotic stress increases, the number of species decreases. This means that species are more likely to become population threatened, endangered, and even extinct, when and where abiotic stress is especially harsh.

ABIOTIC COMPONENT

In biology and ecology, abiotic components or abiotic factors are non-living chemical and physical parts of the environment that affect living organisms and the functioning of ecosystems. Abiotic factors and phenomena associated with them underpin all biology.

Abiotic components include physical conditions and non-living resources that affect living organisms in terms of growth, maintenance, and reproduction. Resources are distinguished as substances or objects in the environment required by one organism and consumed or otherwise made unavailable for use by other organisms.

Component degradation of a substance occurs by chemical or physical processes, e.g. hydrolysis. All non-living components of an ecosystem, such as the atmosphere or water, are called abiotic components.

Examples

In biology, abiotic factors can include water, light, radiation, temperature, humidity, atmosphere, and soil. The macroscopic climate often influences each of the above. Pressure and sound waves may also be considered in the context of marine or sub-terrestrial environments.

All of these factors affect different organisms to different extents. If there is little or no sunlight then plants may wither and die from not being able to get enough sunlight to complete the cycle of photosynthesis. Many Archea require very high temperatures, or pressures, or unusual concentrations of chemical substances, such as sulfur, because of their specialization into extreme conditions. Certain fungi have evolved to survive mostly at the temperature, the humidity, and stability of their environment.

For example, there is a significant difference in access to water as well as humidity between temperate rain forests and deserts. This difference in water access causes a diversity in the types of plants and animals that grow in these areas.

REFERENCES

1. "Abiotic Stress". Biology Online. Archived from the original on 13 June 2008. Retrieved 2008-05-04.
2. Vinebrooke, Rolf D. et al. "Impacts of multiple stressors on biodiversity and ecosystem functioning: the role of species co-tolerance." OIKOS 104: 451–457, 2004.
3. Gao, Ji-Ping, et al. "Understanding Abiotic Stress Tolerance Mechanisms: Recent Studies on Stress Response in Rice." Journal of Integrative Plant Biology 49 (6): 742–750, 2007.
4. Mittler, Ron. "Abiotic stress, the field environment and stress combination." Trends in Plant Science 11(1): 15–19, 2006.
5. Palta, Jiwan P. and Farag, Karim. "Methohasds for enhancing plant health, protecting plants from biotic and abiotic stress related injuries and enhancing the recovery of plants injured as a result of such stresses." United States Patent 7101828, September 2006.
6. Wolfe, A. "Patterns of biodiversity." Ohio State University, 2007.
7. Brussaard, Lijbert, Peter C. de Ruiter, and George G. Brown. "Soil biodiversity for agricultural sustainability." Agriculture, Ecosystems and Environment 121: 233–244, 2007.
8. Roelofs, D. et al. "Functional ecological genomics to demonstrate general and specific responses to abiotic stress." Functional Ecology 22: 8–18, 2008.
9. Wang, W., Vinocur, B. and Altman, A. "Plant responses to drought, salinity and extreme temperatures towards genetic engineering for stress tolerance." Planta 218: 1-14, 2007.

10. Maestre, Fernando T., Jordi Cortina, and Susana Bautista. "Mechanisms underlying the interaction between Pinus halepensis and the native late-successional shrub Pistacia lentiscus in a semi-arid plantation." Ecography 27: 776–786, 2007.
11. Palm, Brady; Van Volkenburgh (2012). "Serpentine tolerance in Mimuslus guttatus does not rely on exclusion of magnesium". Functional Plant Biology.
12. Savvides, Andreas (December 15, 2015). "Chemical Priming of Plants Against Multiple Abiotic Stresses: Mission Possible?". Trends in Plant Science. Retrieved March 10, 2016.
13. Raghothama, K. G. (1999-01-01). "Phosphate Acquisition". Annual Review of Plant Physiology and Plant Molecular Biology. 50 (1): 665–693. doi:10.1146/annurev.arplant.50.1.665. PMID 15012223.
14. Rubio, Vicente; Linhares, Francisco; Solano, Roberto; Martín, Ana C.; Iglesias, Joaquín; Leyva, Antonio; Paz-Ares, Javier (2001-08-15). "A conserved MYB transcription factor involved in phosphate starvation signaling both in vascular plants and in unicellular algae". Genes & Development. 15 (16): 2122–2133. doi:10.1101/gad.204401. ISSN 0890-9369. PMC 312755 . PMID 11511543.
15. Pant, Bikram Datt; Burgos, Asdrubal; Pant, Pooja; Cuadros-Inostroza, Alvaro; Willmitzer, Lothar; Scheible, Wolf-Rüdiger (2015-04-01). "The transcription factor PHR1 regulates lipid remodeling and triacylglycerol accumulation in Arabidopsis thaliana during phosphorus starvation". Journal of Experimental Botany. 66 (7): 1907–1918. doi:10.1093/jxb/eru535. ISSN 0022-0957. PMC 4378627 . PMID 25680792.
16. Pant, Bikram-Datt; Pant, Pooja; Erban, Alexander; Huhman, David; Kopka, Joachim; Scheible, Wolf-Rüdiger (2015-01-01). "Identification of primary and secondary metabolites with phosphorus status-dependent abundance in Arabidopsis, and of the transcription factor PHR1 as a major regulator of metabolic changes during phosphorus limitation". Plant, Cell & Environment. 38 (1): 172–187. doi:10.1111/pce.12378. ISSN 1365-3040.
17. Goncalves-Alvim, Silmary J. and G. Wilson Fernandez. "Biodiversity of galling insects: historical, community and habitat effects in four neotropical savannas." Biodiversity and Conservation 10: 79–98, 2001.
18. Ricklefs, R.E. 2005. The Economy of Nature, 6th edition. WH Freeman, USA.
19. Chapin, F.S. III, H.A. Mooney, M.C. Chapin, and P. Matson. 2011. Principles of terrestrial ecosystem ecology. Springer, New York.
20. Water Quality Vocabulary. ISO 6107-6:1994.
21. Hogan, C. Benito (2010). "Abiotic factor". Encyclopedia of Earth. Washington,D.C.: National Council for Science and the Environment.
22. "Abiotic Components". Department of Biodiversity and Conservation Biology, University of the Western Cape

CHAPTER 3

Heat Stress Tolerance

Chapter Contents

- **Introduction**
- **Breeding For Heat Stress Tolerance In Plants**
- **Traits Associated With Heat Stress Tolerance**

Heat stress is one of the main abiotic stresses that affect the production of various crops in many parts of the world. High temperatures alter several metabolic processes reducing photosynthesis activity that results mainly in grain yield losses. In this context, identifying heat-stress tolerant genotypes with high yield potential could contribute to increasing food production. This chapter deals with aspects of breeding for heat-stress tolerance. First, heat stress and the tolerance mechanisms are characterized along with the main characteristics associated with heat stress, such as cell membrane thermal stability, canopy temperature depression, and some morphological characteristics. It then presents genetic control and selection environments for heat-stress tolerance, breeding methods, and selection strategies. The chapter ends with information on the use of biotechnological tools applied to breeding for heat-stress tolerance.

INTRODUCTION

Plant breeding is process of development of new cultivars. Plant breeding involves development of varieties for different environmental conditions – some of them are not favorable. Among them, heat stress is one of such factor that reduces the production and quality significantly. So breeding against heat is a very important criterion for breeding for current as well as future environments produced by global climate change (e.g. global warming).

BREEDING FOR HEAT STRESS TOLERANCE IN PLANTS

Heat stress due to increased temperature is a very important problem globally. Occasional or prolonged high temperatures cause different morpho-anatomical, physiological and biochemical changes in plants. The ultimate effect is on plant growth as well as development and reduced yield and quality. Breeding for heat stress tolerance can be mitigated by breeding plant varieties that have improved levels of thermo-tolerance using different conventional or advanced genetic tools.

What is Heat Stress Tolerance

Heat stress is defined as increased temperature level sufficient to cause irreversible damage to plant growth and development. Generally a temperature rise, above usually 10 to 15 °C above ambient, can be considered heat shock or heat stress. Heat tolerance is broadly defined as the ability of the plant tolerate heat – means that grow and produce economic yield under high temperatures.

Significance: Current and Future - Global Warming

Heat stress is a serious threat to crop production globally (Hall, 2001, 1992). Global warming is particularly consequence of increased level of green house gases such as CO_2, methane, chlorofluorocarbons and nitrous oxides. The Intergovernmental Panel on Climatic Change (IPCC) has predicted a rise of 0.3 °C per decade (Jones et al., 1999) reaching to approximately 1 and 3 °C above the present value by 2025 and 2100 AD, respectively.

Physiological Consequence of Heat Stress

Extremely high temperatures can cause severe cellular injury, which ultimately lead to the collapse og cellular organization and cell death within a short period of time (Schoffl et al, 1999). Under moderately high temperatures, injury can only occur after longer exposure to such temperature; however the plant efficiency can be still be affected. High temperature directly affect injuries such as protein denaturation and aggregation, and increased fluidity of membrane lipids. Other indirect or slower heat injuries involve inactivation of enzymes in chloroplast and mitochondria, protein degradation, inhibition of protein synthesis, and loss of membrane integrity. Heat stress associated injuries ultimately lead to starvation, inhibition of growth, reduced ion flux, production of toxic compounds and production of reactive oxygen species (ROS). Immediately after exposure to high temperature stress-related proteins are expressed as stress defense strategy of the cell. Expression of heat shock proteins (HSPs), protein with 10 to 200 kDa, is supposed to be involved in signal transduction during heat stress. In many species it has been demonstrated that HSPs results in improved physiological phenomena such as photosynthesis, assimilate partitioning, water and nutrient use efficiency, and membrane stability.

Studies have found tremendous variation within and between species, thus this will help to breed heat tolerance for future environment. Some of attempts to develop heat-tolerant genotypes are successful. (Ehlers and Hall, 1998; Camejo et al., 2005)

TRAITS ASSOCIATED WITH HEAT STRESS TOLERANCE

Different physiological mechanisms , such as heat tolerant metabolism as indicated by higher photosynthetic rates, stay-green, membrane thermo-stability, and heat avoidance as indicated by canopy temperature depression, may contribute to heat tolerance in the field. Several physiological and morphological traits have been evaluated for heat tolerance - Canopy temperature, leaf chlorophyll, stay green, leaf conductance, spike number, biomass, and flowering date.

(a) Canopy temperature depression (CTD)

CTD has shown clear association with yield in warm environments shows it association with heat stress tolerance. CTD shows high genetic correlation with yield and high values of proportion of direct response to selection (Reynolds et al., 1998)indicating that the trait is heritable and therefore amenable to early generation selection. Since an integrated CTD value can be measured almost instantaneously on scores of plants in a small breeding plot (thus reducing error normally associated with traits measured on individual plants), work has been conducted to evaluate its potential as an indirect selection criterion for genetic gains in yield. CTD is affected by many physiological factors, which makes it a powerful.

(b) Stomatal conductance

Canopy temperature depression is highly suitable for selecting physiologically superior lines in warm, low relative humidity environments where high evaporative demand leads to leaf cooling of up to 10 °C below ambient temperatures. This permits differences among genotypes to be detected relatively easily using infrared thermometry. However, such differences cannot be detected in high relative humidity environments because the effect of evaporative cooling of leaves is negligible. Nonetheless, leaves maintain their stomata open to permit the uptake of CO2, and differences in the rate of CO2 fixation may lead to differences in leaf conductance that can be measured using a porometer. Porometry can be used to screen individual plants. The heritability of stomatal conductance is reasonably high, with reported values typically in the range of 0.5 to 0.8. Plants can be assessed for leaf conductance using a viscous flow porometer that is available on the market (Thermoline and CSIRO, Australia). This instrument can give a relative measure of stomatal conductance in a few seconds, making it possible to identify physiologically superior genotypes from within bulks.

(C) Membrane thermostability

Although resistance to high temperatures involves several complex tolerance and avoidance mechanisms, the membrane is thought to be a site of primary physiological injury by heat, and measurement of solute leakage from tissue can be used to estimate damage to membranes. Since membrane thermostability is reasonably heritable (Fokar et al., 1998) and shows high genetic correlation with yield.

(D) Chlorophyll fluorescence

Chlorophyll fluorescence, an indication of the fate of excitation energy in the photosynthetic apparatus, has been used indicator for heat stress tolerance.

(E) Chlorophyll content and stay green

Chlorophyll content and stay green traits have been found to be associated with heat stress tolerance.,. Xu et al. (2000) identified three QTLs for chlorophyll content (Chl1, Chl2, and Chl3) (coincided with three stay-green QTL regions (Stg1, Stg2, and Stg3)) were identified in Sorghum. The Stg1 and Stg2 regions also contain the genes for key photosynthetic enzymes, heat shock proteins, and an abscisic acid (ABA) responsive gene.

(F) Photosynthesis

Decliend photosynthesis is suggested as measure of heat stress sensitivity in plants.

(G) Stem reserve remobilization'

Combination Breeding and Physiological Breeding

The physiological-trait-based breeding approach has merit over breeding for yield *per se* because it increases the probability of crosses resulting in additive gene action. The concept of combination phenomics comes from the idea that two or more stress have common physiological effect or common traits - which is an indicator of overall plant health. Similar analogy in human medical terms is high blood pressure or high body temperature or high white blood cells in body is an indicator of health problems and thus we can select healthy people from unhealthy using such a measure. As both abiotic and abiotic stresses can result in similar physiological consequence, tolerant plant can be separated from sensitive plants. Some imaging or infrared measuring techniques can help to speed the process for breeding process. For example, spot blotch intensity and canopy temperature depression can be monitored with canopy temperature depression.

REFERENCES

1. Hall, A.E., 1992. Breeding for heat tolerance. Plant Breed. Rev. 10, 129–168.
2. Hall, A.E., 2001. Crop Responses to Environment. CRC Press LLC, Boca Raton, Florida
3. Jones, P.D., New, M., Parker, D.E., Mortin, S., Rigor, I.G., 1999. Surface area temperature and its change over the past 150 years. Rev. Geophys. 37, 173–199
4. Schoffl, F., Prandl, R., Reindl, A., 1999. Molecular responses to heat stress. In: Shinozaki, K., Yamaguchi-Shinozaki, K. (Eds.), Molecular Responses to Cold, Drought, Heat and Salt Stress in Higher Plants. R. G. Landes Co.,Austin, Texas, pp. 81–98
5. Ehlers, J.D., Hall, A.E., 1998. Heat tolerance of contrasting cowpea lines in short and long days. Field Crops Res. 55, 11–21.

6. Camejo, D., Jimenez, A., Alarc ′ on, J.J., Torres, W., G ′ omez, J.M., Sevilla, ′F., 2006. Changes in photosynthetic parameters and antioxidant activities following heat-shock treatment in tomato plants. Funct. Plant Biol. 33,177–18

7. Reynolds, M.P., R.P. Singh, A. Ibrahim, O.A.A. Ageeb, A. Larqué-Saavedra, and J.S. Quick.1998. Evaluating physiological traits to complement empirical selection for wheat in warm environments. Euphytica 100:84-95.,

8. Fokar, M., H.T. Nguyen, and A. Blum. 1998. Heat tolerance in spring wheat. I. Genetic variability and heritability of cellular thermotolerance. Euphytica 104:1-8.

9. Rosyara, U. R., S. Subdedi, R. C. Sharma and E. Duveiller. 2010. Photochemical Efficiency and SPAD Value as Indirect Selection Criteria for Combined Selection of Spot Blotch and Terminal Heat Stress in Wheat. Journal of Phytopathology Volume 158, Issue 11-12, pages 813–821, December 2010

10. Rosyara, U. R., S. Subdedi, R. C. Sharma and E. Duveiller.2010.The effect of spot blotch and heat stress in variation of canopy temperature depression, chlorophyll fluorescence and chlorophyll content of hexaploid wheat genotypes. Euphytica Volume 174, Number 3, 377-390

11. Rosyara, U. R., S. Subdedi, R. C. Sharma and E. Duveiller. 2010. Photochemical Efficiency and SPAD Value as Indirect Selection Criteria for Combined Selection of Spot Blotch and Terminal Heat Stress in Wheat. Journal of Phytopathology Volume 158, Issue 11-12, pages 813–821, December 2010

12. Xu W, Subudhi PK, Crasta OR, Rosenow DT, Mullet JE, Nguyen HT. Molecular mapping of QTLs conferring stay-green in grain sorghum (Sorghum bicolor L. Moench). Genome. 2000 Jun;43(3):461-9.

13. Berry J. and O. Bjorkman. 1980. Photosynthetic response and adaptation to temperature in higher plants. Annu. Rev. Plant Physiol. 31:491–543.

14. Physiological approaches to wheat breeding, http://www.fao.org/docrep/006/y4011e/y4011e0a.htm

15. U.R. Rosyara, N.K. Gupta, S. Gupta, R.C. Sharma (2011) Genomics and Phenomics of Chlorophyll Associated Traits in Abiotic Stress Tolerance Breeding In: Hua Le and Elisa Salcedo, Chlorophyll: Structure, Production and Medicinal Uses, N.Y. : Nova Science Publishers

16. Rosyara U. R., Gupta N. K., Gupta S., R.C. Sharma (2012) Breeding for Heat Stress Tolerance in Plants via Molecular and Physiological Manipulation In: Stanislas Josipovic and Elias Ludwig, Heat Stress: Causes, Treatment and Prevention, N.Y. : Nova Science Publishers

17. Rosyara, U.R., Vromman, D., Duveiller, E. 2008. Canopy temperature depression as indication of correlative measure of spot blotch resistance and heat stress tolerance in spring wheat. J. Plant Path. 90 :103–107.

18. Rosyara, U. R., S. Subdedi, R. C. Sharma and E. Duveiller.2010.The effect of spot blotch and heat stress in variation of canopy temperature depression, chlorophyll fluorescence and chlorophyll content of hexaploid wheat genotypes. Euphytica Volume 174, Number 3, 377-390

19. Rosyara, U. R., S. Subdedi, R. C. Sharma and E. Duveiller. 2010. Photochemical Efficiency and SPAD Value as Indirect Selection Criteria for Combined Selection of Spot Blotch and Terminal Heat Stress in Wheat. Journal of Phytopathology Volume 158, Issue 11-12, pages 813–821

20. Rosyara, U.R., Vromman, D., Duveiller, E. 2008. Canopy temperature depression as indication of correlative measure of spot blotch resistance and heat stress tolerance in spring wheat. J. Plant Path. 90 :103–107.

CHAPTER 4

Hardiness of Plants

Chapter Contents

Plant hardiness is often a misunderstood topic that deserves special discussion. It is not as simple as it may appear. All factors involved must be taken into account when determining the hardiness of a plant, in order to avoid disappointment if the plant wilts or does not bloom.

INTRODUCTION

Hardiness of plants describes their ability to survive adverse growing conditions. It is usually limited to discussions of climatic adversity. Thus a plant's ability to tolerate cold, heat, drought, flooding, or wind are typically considered measurements of hardiness. Hardiness of plants is defined by their native extent's geographic location: longitude, latitude and elevation. These attributes are often simplified to a hardiness zone. In temperate latitudes, the term most often describes resistance to cold, or "cold-hardiness", and is generally measured by the lowest temperature a plant can withstand. Hardiness of a plant is usually divided into two categories: tender, and hardy. Some sources also use the erroneous terms "Half-hardy" or "Fully hardy". Tender plants are those killed by freezing temperatures, while hardy plants survive freezing—at least down to certain temperatures, depending on the plant. "Half-hardy" is a term used sometimes in horticulture to describe bedding plants which are sown in heat in winter or early spring, and planted outside after all danger of frost has passed. "Fully hardy" usually refers to plants being classified under the Royal Horticultural Society classifications, and can often cause confusion to those not using this method.

Plants vary a lot in their tolerance of growing conditions. The selective breeding of varieties capable of withstanding particular climates forms an important part of agriculture and horticulture. Plants adapt to changes in climate on their own to some extent. Part of the work of nursery growers of plants consists of cold hardening, or hardening off their plants, to prepare them for likely conditions in later life.

COLD HARDENING

Cold hardening is the physiological and biochemical process by which an organism prepares for cold weather.

Plants

Plant covered in snow after an ice storm in 2013, Ontario, Canada

Plants in temperate and polar regions adapt to winter and sub zero temperatures by relocating nutrients from leaves and shoots to storage organs. Freezing temperatures induce dehydrative stress on plants, as water absorption in the root and water transport in the plant decreases. Water in and between cells in the plant freezes and expands, causing tissue damage. Cold hardening is a process in which a plant undergoes physiological changes to avoid, or mitigate cellular injuries caused by sub-zero temperatures. Non-acclimatized individuals can survive -5 °C, while a acclimatized individual in the same species can survive -30°. Plants that originated in the tropics, like tomato or maize, don't go through cold hardening and are unable to survive freezing temperatures. The plant starts the adaptation by exposure to cold yet still not freezing temperatures. The process can be divided into three steps. First the plant perceives low temperature, then converts the signal to activate or repress expression of appropriate genes. Finally, it uses these genes to combat the stress, caused by sub-zero temperatures, affecting its living cells. Many of the genes and responses to low temperature stress are shared with other abiotic stresses, like drought or salinity.

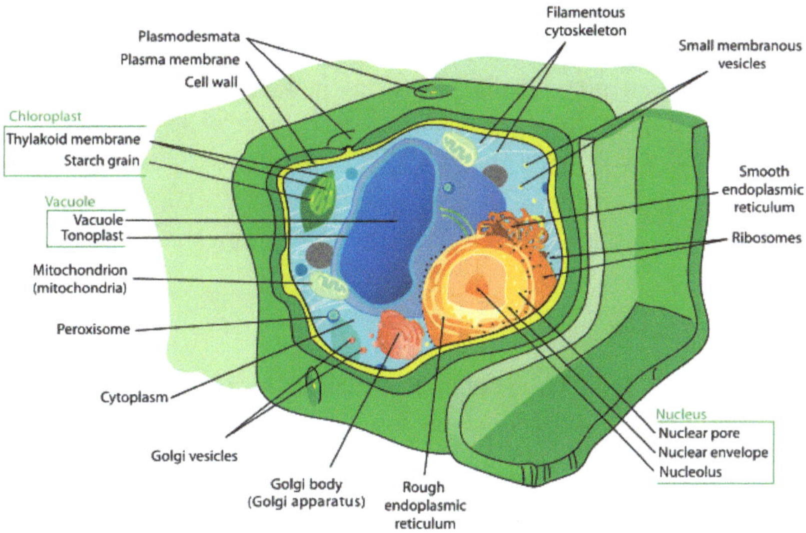

Schematic of typical plant cell

When temperature drops, the membrane fluidity, RNA and DNA stability, and enzyme activity change. These, in turn, affect transcription, translation, intermediate metabolism, and photosynthesis, leading to an energy imbalance. This energy imbalance is thought to be one of the ways the plant detects low temperature. Experiments on *arabidopsis* show that the plant detects the change in temperature, rather than the absolute temperature.The rate of temperature drop is directly connected to the magnitude of calcium influx, from the space between cells, into the cell. Calcium channels in the cell membrane detect the temperature drop, and promotes expression of low temperature responsible genes in *alfalfa* and *arabidopsis*. The response to the change in calcium elevation depends on the cell type and stress history. Shoot tissue will respond more than a root cells, and a cell that already is adapted to cold stress will respond more than one that has not been through cold hardening before. Light doesn't control the onset of cold hardening directly, but shortening of daylight is associated with fall, and starts production of reactive oxygen species and excitation of photosystem 2, which influences low-temp signal transduction mechanisms. Plants with compromised perception of day length have compromised cold acclimation.

Cold increases cell membrane permeability and makes the cell shrink, as water is drawn out when ice is formed in the extracellular matrix between cells. To retain the surface area of the cell membrane so it will be able to regain its former volume when temperature rises again, the plant forms more and stronger Hechtian strands. These are tubelike structures that connect the protoplast with the cell wall. When the intracellular water freezes, the cell will expand, and without cold hardening the cell would rupture. To protect the cell membrane from expansion induced damage, the plant cell changes the proportions of almost all lipids in the cell membrane, and increases the amount of total soluble protein and other cryoprotecting molecules, like sugar and proline.

Chilling injury occurs at 0–10 degrees Celsius, as a result of membrane damage, metabolic changes, and toxic buildup. Symptoms include wilting, water soaking, necrosis, chlorosis, ion leakage, and decreased growth. Freezing injury may occur at temperatures below 0 degrees Celsius. Symptoms of extracellular freezing include structural damage, dehydration, and necrosis. If intracellular freezing occurs, it will lead to death. Freezing injury is a result of lost permeability, plasmolysis, and post-thaw cell bursting.

When spring comes, or during a mild spell in winter, plants de-harden, and if the temperature is warm for long enough – their growth resumes.

Insects

Cold hardening has also been observed in insects such as the fruit fly and diamondback moth. The insects use rapid cold hardening to protect against cold shock during overwintering periods.Overwintering insects stay awake and active through the winter while non-overwintering insects migrate or die. Rapid cold hardening can be experienced during short periods of undesirable temperatures, such as cold shock in environment temperature, as well as the common cold months. The build up of cryoprotective compounds is the reason that insects can experience cold hardening. Glycerol is a cryoprotective substance found within these insects capable of overwintering. Through testing, glycerol requires interactions with other cell components within the insect in order to decrease the body's permeability to the cold. When an insect is exposed to these cold temperatures, glycerol rapidly accumulates. Glycerol is known as a non-ionic kosmotropeforming powerful hydrogen bonds with water molecules. The hydrogen bonds in the glycerol compound compete with the weaker bonds between the water molecules causing an interruption in the makeup of ice formation. This chemistry found within the glycerol compound and reaction between water has been used as an antifreeze in the past, and can be seen here when concerning cold hardening. Proteins also play a large role in the cryoprotective compounds that increase ability to survive the cold hardening process and environmental change. Glycogen phosphorylase(GlyP) has been a key protein found during testing to increase in comparison to a controlled group not experiencing the cold hardening. Once warmer temperatures are observed the process of acclimation begins, and the increased glycerol along with other cryoprotective compounds and proteins are also reversed. There is a rapid cold hardening capacity found within certain insects that suggests not all insects can survive a long period of overwintering. Non-diapausing insects can sustain brief temperature shocks but often have a limit to which they can handle before the body can no longer produce enough cryoprotective components.

The common fruit fly

Inclusive to the cold hardening Aside from being beneficial for the survival of an insect during cold temperatures, the cold hardening process also helps improve the performance of the organism Rapid cold hardening (RCH) is one of the fastest cold temperature responses recorded. This process allows an insect to instantly adapt to the severe weather change without compromising function. The *Drosophila melanogaster* (common fruit fly) is a frequently experimented insect involving cold hardening. A proven example of RCH enhancing performance comes from courting and mating within the fruit fly. It has been tested that the fruitflies which mated more frequently have enhanced RCH than the control group. Most insects experiencing extended cold periods are observed to modify the membrane lipids within the body. Desaturation of fatty acids are the most commonly seen modification to the membrane. When the fruit fly was observed under the stressful climate the survival rate increased in comparison to the fly prior to cold hardening.

The diamondback moth

In addition to testing on the common fruit fly, *Plutella xylostella* (diamondback moth) also has been widely studied for its significance in cold hardening. While this insect also shows an increase in glycerol and similar cryoprotective compounds, it also shows an increase in polyols. These compounds are specifically linked to cryoprotective compounds designed to withstand cold hardening. The polyol compound is freeze-susceptible and freeze tolerant. Polyols simply act as a barrier within the insect body by preventing intracellular freezing by restricting the extracellular freezing likely to happen in overwintering periods. During the larvae stage of the diamondback moth, the significance of glycerol was tested again for validity. The lab injected the larvae with added glycerol and in turn proved that glycerol is a major factor in survival rate when cold hardening. The cold tolerance is directly proportional to the build up of glycerol during cold hardening.

Cold hardening of insects improves the survival rate of the species and improves function. Once environmental temperature begins to warm up above freezing, the cold hardening process is reversed and the glycerol and cryprotective compounds decrease within the body. This also reverts the function of the insect to pre-cold hardening activity.

HARDINESS ZONE

A hardiness zone (a subcategory of vertical zonation) is a geographically defined area in which a specific category of plant life is capable of growing, as defined by climatic conditions, including its ability to withstand the minimum temperatures of the zone (see the scale on the right or the table below). For example, a plant that is described as "hardy to zone 10" means that the plant can withstand a minimum temperature of −1 °C (30 °F). A more resilient plant that is "hardy to zone 9" can tolerate a minimum temperature of −7 °C (19 °F). First developed by the United States Department of Agriculture (USDA), the use of the zones has been adopted by other countries.

Usda Hardiness Zones

Zone		From	To
0	a	< −53.9 °C (−65 °F)	
	b	−53.9 °C (−65 °F)	−51.1 °C (−60 °F)
1	a	−51.1 °C (−60 °F)	−48.3 °C (−55 °F)
	b	−48.3 °C (−55 °F)	−45.6 °C (−50 °F)
2	a	−45.6 °C (−50 °F)	−42.8 °C (−45 °F)
	b	−42.8 °C (−45 °F)	−40 °C (−40 °F)
3	a	−40 °C (−40 °F)	−37.2 °C (−35 °F)
	b	−37.2 °C (−35 °F)	−34.4 °C (−30 °F)
4	a	−34.4 °C (−30 °F)	−31.7 °C (−25 °F)
	b	−31.7 °C (−25 °F)	−28.9 °C (−20 °F)
5	a	−28.9 °C (−20 °F)	−26.1 °C (−15 °F)
	b	−26.1 °C (−15 °F)	−23.3 °C (−10 °F)
6	a	−23.3 °C (−10 °F)	−20.6 °C (−5 °F)
	b	−20.6 °C (−5 °F)	−17.8 °C (0 °F)

7	a	−17.8 °C (0 °F)	−15 °C (5 °F)
	b	−15 °C (5 °F)	−12.2 °C (10 °F)
8	a	−12.2 °C (10 °F)	−9.4 °C (15 °F)
	b	−9.4 °C (15 °F)	−6.7 °C (20 °F)
9	a	−6.7 °C (20 °F)	−3.9 °C (25 °F)
	b	−3.9 °C (25 °F)	−1.1 °C (30 °F)
10	a	−1.1 °C (30 °F)	+1.7 °C (35 °F)
	b	+1.7 °C (35 °F)	+4.4 °C (40 °F)
11	a	+4.4 °C (40 °F)	+7.2 °C (45 °F)
	b	+7.2 °C (45 °F)	+10 °C (50 °F)
12	a	+10 °C (50 °F)	+12.8 °C (55 °F)
	b	> +12.8 °C (55 °F)	

Benefits and Drawbacks

The hardiness zones are informative: the extremes of winter cold are a major determinant of whether a plant species can be cultivated outdoors at a particular location; however, the USDA hardiness zones have a number of drawbacks if used without supplementary information. The zones do not incorporate summer heat levels into the zone determination; thus sites which may have the same mean winter minima, but markedly different summer temperatures, will be accorded the same hardiness zone.

Another issue is that the hardiness zones do not take into account the reliability of the snow cover. Snow acts as an insulator against extreme cold, protecting the root system of hibernating plants. If the snow cover is reliable, the actual temperature to which the roots are exposed will not be as low as the hardiness zone number would indicate. As an example, Quebec City in Canada is located in zone 4, but can rely on a significant snow cover every year, making it possible to cultivate plants normally rated for zones 5 or 6. But, in Montreal, located to the southwest in zone 5, it is sometimes difficult to cultivate plants adapted to the zone because of the unreliable snow cover.

Other factors that affect plant survival, though not considered in hardiness zones, are soil moisture, humidity, the number of days of frost, and the risk of a rare catastrophic cold snap. Some risk evaluation – the probability of getting a particularly severe low temperature – often would be more useful than just the average conditions.

Lastly, many plants may survive in a locality but will not flower if the day length is insufficient or if they require vernalization (a particular duration of low temperature). With annuals, the time of planting can often be adjusted to allow growth beyond their normal geographical range.

An alternative means of describing plant hardiness is to use "indicator plants" (the USDA also publishes a list of these to go with the hardiness zone map). In this method, common plants with known limits to their range are used.

Gardening books are available that provide more information on climate zones. For example, Sunset Books (associated with *Sunset* magazine) publishes a series that breaks up climate zones more finely than the USDA zones. They identify 45 distinct zones in the US, incorporating ranges of temperatures in all seasons, precipitation, wind patterns, elevation, and length and structure of the growing season.

European Hardiness Zones

Britain and Ireland

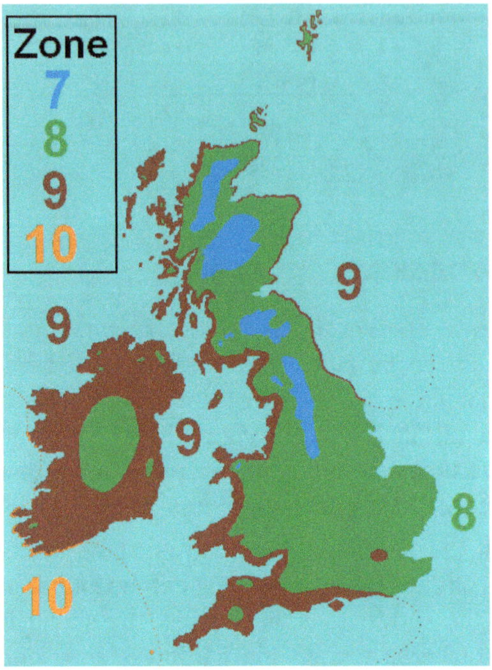

Britain and Ireland's hardiness zones

Owing to the moderating effect of the North Atlantic Current on the Irish and British temperate maritime climate, Britain, and Ireland even more so, have milder winters than their northerly position would otherwise afford. This means that the hardiness zones relevant to Britain and Ireland are quite high, from 7 to 10, as shown below.

- 7. In Scotland the Grampians, Highlands and locally in the Southern Uplands, in England the Pennines and in Wales the highest part of Snowdonia.
- 8. Most of England, Wales and Scotland, parts of central Ireland, and Snaefell on the Isle of Man.
- 9. Most of western and southern England and Wales, western Scotland, also a very narrow coastal fringe on the east coast of Scotland and northeast England (within 5 km of the North Sea), London, most of Ireland, and most of the Isle of Man.
- 10. Very low-lying coastal areas of the southwest of Ireland and the Isles of Scilly.

In 2012 the United Kingdom's Royal Horticultural Society introduced new hardiness ratings for plants, from H7, the hardiest (tolerant of temperatures below -20°C) to H1a (needing temperatures above 15°C).

Northern Europe

Scandinavia lies at the same latitude as Alaska or Greenland, but the effect of the warm North Atlantic Current is even more pronounced here than it is in Britain and Ireland. Save for a very small spot near Karasjok, Norway, which is in zone 2, nowhere in the Arctic part of Scandinavia does it get below zone 3. The Faroe Islands, at 62–63°N are in zone 8, as are the outer Lofoten Islands at 68°N. Tromsø, a coastal city in Norway at 70°N, is in zone 7, and even Longyearbyen, the northernmost true city in the world at 78°N, is still in zone 5. All these coastal locations have one thing in common, though, which are cold, damp summers, with temperatures rarely exceeding 20 °C (68 °F), or 15 °C (59 °F) in Longyearbyen. This shows the importance of taking heat zones into account for better understanding of what may or may not grow.

A garden in Simrishamn, southern Sweden.

In Sweden and Finland generally, at sea level to 500 metres (1,600 ft), zone 3 is north of the Arctic Circle, including cities like Karesuando, Pajala and Rovaniemi. Kiruna is the major exception here, which being located on a hill above frost traps, is in zone 5. Zone 4 lies between the Arctic Circle and about 64–65°N, with cities such as Oulu and Jokkmokk, zone 5 (south to 61–62°N) contains cities such as Tampere, Umeå and Östersund. Zone 6 covers the south of mainland Finland, Sweden north of 60°N, and the high plateau of Småland further south. Here one will find cities such as Gävle, Örebro, Sundsvall and Helsinki. The Åland Islands, as well as coastal Southern Sweden, and the Stockholm area are in zone 7. The west coast of Sweden (Gothenburg and southwards) enjoys particularly mild winters and lies in zone 7, therefore being friendly to some hardy exotic species (found, for example, in the Gothenburg Botanical Garden), the southeast coast of Sweden has a colder winter due to the absence of the Gulf Stream.

Central Europe

Central Europe is a good example of a transition from an oceanic climate to a continental climate, which can be noticed immediately when looking at the hardiness zones, which tend to decrease mainly eastwards instead of northwards. Also, the plateaux and low mountain ranges in this region have a significant impact on how cold it might get during winter. Generally speaking, the hardiness zones are high considering the latitude of the region, although not as high as in the Shetland Islands where zone 9 extends to over 60°N. In Central Europe, the relevant zones decrease from zone 8 on the Belgian, Dutch and German North Sea coast, with the exception of some of the Frisian

Islands (notably Vlieland and Terschelling), the island of Heligoland and some of the islands in the Rhine-Scheldt estuary, which are in zone 9, to zone 5 around Suwałki, Podlachia on the far eastern border between Poland and Lithuania. Some isolated, high elevation areas of the Alps and Carpathians may even go down to zone 3 or 4. An extreme example of a cold sink is Funtensee, Bavaria which is at least in zone 3 and maybe even in zone 1 or 2. Another notable example is Waksmund, a small village in the Polish Carpathians, which regularly reaches −35 °C (−31 °F) during winter on calm nights when cold and heavy airmasses from the surrounding Gorce and Tatra Mountains descend down the slopes to this low-lying valley, creating extremes which can be up to 10 °C (18 °F) colder than nearby Nowy Targ or Białka Tatrzańska, which are both higher up in elevation. Waksmund is in zone 3b while nearby Kraków, only 80 km (50 mi) to the north and 300 m (980 ft) lower is in zone 6a. These examples prove that local topography can have a pronounced effect on temperature and thus on what is possible to grow in a specific region.

Southern Europe

In Southern Europe, the marker plant for climate, which also serves as a cultural indicator, is the olive tree. It cannot survive long periods with temperature below zero, so its cultivation area matches the cool winter zone. The mediterranean sea acts as a temperature regulator so this area is generally warmer than other parts of the continent; except in mountainous areas where the sea effect lowers, it belongs to 8–10 zones; even, southern Balkans (mountainous-Western and Eastern Serbia, continental Croatia and Bulgaria) are colder in winter and are in zones 6–7. The Croatian (Dalmatian) coast, Albania and northern Greece are in zones 8–9, as are central-northern Italy (hills and some spots in Po Valley are however colder) and southern France; Central Iberia is 8–9 (some areas are slightly colder). The Spanish and Portuguese Atlantic coast, most of Andalusia and Murcia, almost all the Valencian Community, a part of Catalonia, Balearic Islands, southwestern Sardinia, most of Sicily, coastal southern Italy and southwestern Greece are in zone 10, the islands of Malta, Lampedusa and the southern zone of Spain (Málaga, Cádiz, Granada; where there is mango and avocado cultivation) belong to zone 11a. Also, a few areas on the southernmost coast of Cyprus are inside the zone 11a.

European Cities

The table below provides hardiness zone data for some European cities (based on climatological data):

City	Zone	City	Zone
Alicante, Spain	**10b**	Almería, Spain	**11a**
Amsterdam, The Netherlands	**8b**	Antwerp, Belgium	**8**
Auxerre, France	**8a**	Barcelona, Spain	**10a**
Belfast, Northern Ireland	**9**	Berlin, Germany	**7b**
Bratislava, Slovakia	**7b**	Bucharest, Romania	**6b**
Cardiff, Wales	**9**	Catania, Italy	**10**

Belgrade, Serbia	**8a**	Copenhagen, Denmark	**8a**
Dublin, Ireland	**9a**	Düsseldorf, Germany	**8**
Edinburgh, Scotland	**8b**	Gdańsk, Poland	**7**
Glasgow, Scotland	**8b**	Hamburg, Germany	**8a**
Helsinki, Finland	**6a**	Istanbul, Turkey	**8b**
Kaliningrad, Russia	**6**	Kiev, Ukraine	**6a**
Kraków, Poland	**6**	Lisbon, Portugal	**10b**
Las Palmas, Spain	**12b**	A Coruña, Spain	**10b**
Ljubljana, Slovenia	**7b**	London, England	**9a**
Madrid, Spain	**9a**	Málaga, Spain	**11a**
Marseille, France	**9a**	Milan, Italy	**8b**
Minsk, Belarus	**5b**	Moscow, Russia	**5a**
Munich, Germany	**7b**	Murmansk, Russia	**5**
Nicosia, Cyprus	**10a**	Oslo, Norway	**6a**
Simferopol, Ukraine	**7a**	Palma, Spain	**10a**
Paris, France	**8b**	Poznań, Poland	**7a**
Prague, Czech Republic	**7a**	Reykjavík, Iceland	**8a**
Riga, Latvia	**6a**	Rome, Italy	**9b**
Rovaniemi, Finland	**4**	Saint Petersburg, Russia	**5b**
Sarajevo, Bosnia and Herzegovina	**7b**	Santander, Spain	**10a**
Simrishamn, Sweden	**8a**	Sochi, Russia	**9**
Sofia, Bulgaria	**7a**	Stockholm, Sweden	**6b**
Strasbourg, France	**7**	Tallinn, Estonia	**6b**
Tuapse, Russia	**8**	Tórshavn, Faroe Islands	**7-8**
Tromsø, Norway	**7**	Trondheim, Norway	**6**
Umeå, Sweden	**5**	Valencia, Spain	**10b**
Valletta, Malta	**10b**	Vienna, Austria	**8a**
Vilnius, Lithuania	**5**	Vorkuta, Russia	**2**
Warsaw, Poland	**6b**	Wroclaw, Poland	**7a**
Zagreb, Croatia	**8a**	Zürich, Switzerland	**8a**
Zaragoza, Spain	**8b**		

United States Hardiness Zones

Most of the warmer zones (zones 8, 9, 10, and 11) are located in the southern half of the United States. The low latitude and often stable weather in the far southwestern states, California, and Florida is responsible for the few episodes of severe cold relative to normal in those areas. The more northerly and central portions of the USA are the cooler zones (zones 7, 6, 5, and 4). The central and northerly portions of the mainland often have much less consistent range of temperatures in winter due to being more continental, and thus the zone map has its limitations in these areas.

The USDA first issued its standardized hardiness zone map in 1960, and revised it in 1965. The newest maps were issued in 2012 by the USD, and show a warming of most zones across the USA.

Updates

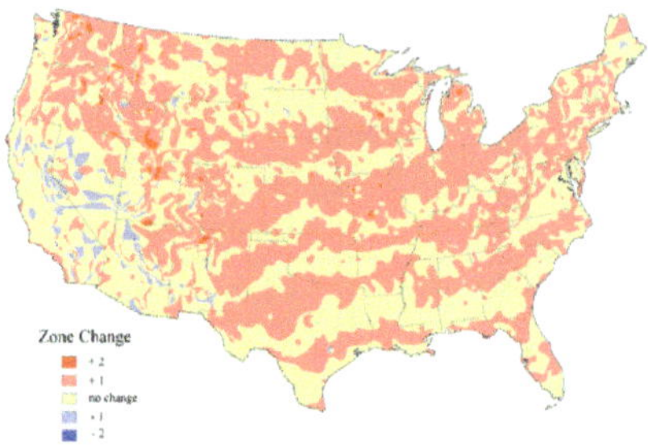

Differences between 1990 USDA hardiness zones and 2006 arborday.org hardiness zones reflect global warming..

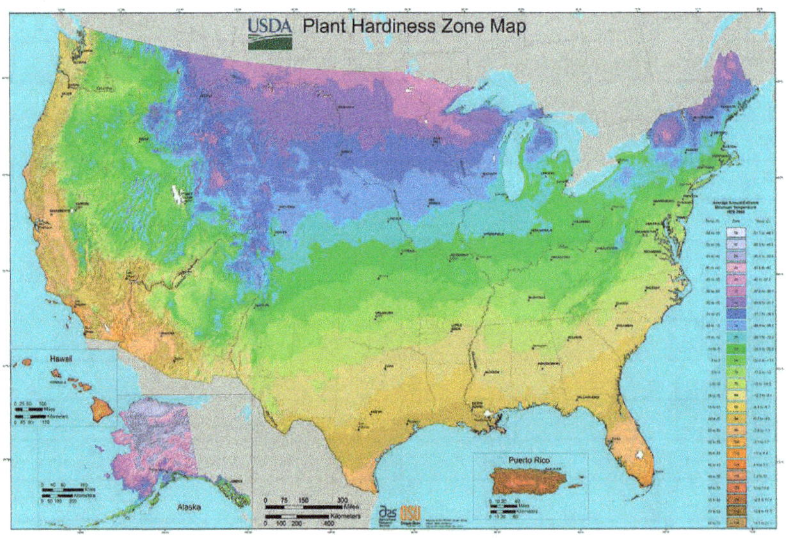

2012 update of the Hardiness Zone Map

In 2003, the American Horticultural Society (AHS) produced a draft revised map, using temperature data collected from July 1986 to March 2002. The 2003 map placed many areas approximately a half-zone higher (warmer) than the 1990 map had. Reviewers noted the map zones appeared to be closer to the original 1960 map in its overall zone delineations. The 2003 AHS draft map purported to show finer detail, for example, reflecting urban heat islands by showing the downtown areas of several cities (e.g., Baltimore, Maryland; Washington, D.C. and Atlantic City, New Jersey) as a full zone warmer than outlying areas. The map excluded the detailed a/b half-zones introduced in the 1990 map, an omission widely criticized by horticulturists and gardeners due to the coarseness of the resulting map. The USDA rejected the AHS 2003 draft map; the agency stated it would create its own map in an interactive computer format. As of August 2010 the AHS and the National Arboretum websites still presented the 1990 map as current.

In 2006, the US National Arbor Day Foundation completed an extensive update of U.S. hardiness zones. It used essentially the same data as the AHS. Once the Foundation analyzed the new data, it revised hardiness zones, reflecting the generally warmer recent temperatures in many parts of the country. The Foundation's 2006 map appears to validate the data used in the AHS 2003 draft. The Foundation also did away with the more detailed a/b half-zone delineations.

In 2012 the USDA updated their plant hardiness map to reflect the warmer observed temperatures in the past thirty years.

U.S. Cities

The USDA plant hardiness zones for U.S. cities as based on the 2012 map are the following:

City	Zone	City	Zone
Albuquerque, New Mexico	**7b**	Oklahoma City, Oklahoma	**7a**
Anchorage, Alaska	**4b/5a**	Omaha, Nebraska	**5b**
Atlanta, Georgia	**7b/8a**	Orlando, Florida	**9b**
Baltimore, Maryland	**7b/8a**	Owensboro, Kentucky	**6b**
Boston, Massachusetts	**6b/7a**	Philadelphia, Pennsylvania	**7a/7b**
Buffalo, New York	**6a**	Phoenix, Arizona	**9b/10a**
Burlington, Vermont	**5a**	Pierre, South Dakota	**4b/5a**
Charleston, West Virginia	**6b**	Pittsburgh, Pennsylvania	**6b**
Chicago, Illinois	**6a**	Portland, Maine	**5b**
Charlotte, North Carolina	**7b/8a**	Portland, Oregon	**8b/9a**
Columbus, Ohio	**6a**	Quad Cities, Iowa/Illinois	**5b**
Dallas, Texas	**8a/8b**	Raleigh, North Carolina	**7b**
Denver, Colorado	**5b/6a**	Reno, Nevada	**6b/7a**
Detroit, Michigan	**6b**	Roanoke, Virginia	**7a/7b**
Fairbanks, Alaska	**2a**	Kansas City, Missouri	**6a/6b**
Hartford, Connecticut	**6b**	Salt Lake City, Utah	**7a/7b**
Honolulu, Hawaii	**12b**	San Antonio, Texas	**8b/9a**
Houston, Texas	**9a**	San Diego, California	**10a/10b**
Indianapolis, Indiana	**5b/6a**		
Las Vegas, Nevada	**9a**	San Francisco, California	**10a/10b**
Los Angeles, California	**10a/10b**	San Jose, California	**9b/10a**
Memphis, Tennessee	**7b/8a**	Savannah, Georgia	**8b**

Miami, Florida	10b/11a	Seattle, Washington	8b
Minneapolis, Minnesota	4b/5a	Tampa, Florida	9b/10a
Nashville, Tennessee	7a	Tucson, Arizona	9a/9b
New Orleans, Louisiana	9b	Tuscaloosa, Alabama	8a
New York, New York	7a/7b	Washington, D.C.	7b/8a
Norfolk, Virginia	8a	Wichita, Kansas	6b
San Juan, Puerto Rico	12b/13a		

Australian Hardiness Zones

The USDA hardiness zones are in use in Australia, but in addition the Australian National Botanic Gardens have devised another system more in keeping with Australian conditions. They are numerically about 7 lower than the USDA system. For example, Australian zone 3 is roughly equivalent to USDA zone 9. The higher Australian zone numbers have no US equivalents.

There are problems with classifications of this type: the spread of weather stations is insufficient to give clear zones and too many places with different climates are lumped together. Only 738 Australian stations have records of more than ten years (one station per 98,491 hectares or 243,380 acres), though more populated areas have relatively fewer hectares per station. Local factors such as aspect, altitude, proximity to the sea also complicate the matter. For example, Mount Isa has three climatic stations with more than a ten-year record. One is in zone 4a, one in zone 4b and the other is in zone 5a. Likewise, Sydney residents can choose between zones 3a and 4b. Most other cities have similar problems. Different locations in the same city are suitable for different plants, making it hard to draw a meaningful map. There may even be a case for publishing a list of weather stations and their zone classification to allow best use of local conditions.

Canadian Hardiness Zones

Climate variables that reflect the capacity and detriments to plant growth are used to develop an index that is mapped to Canada's Plant Hardiness Zones. This index comes from a formula originally developed by Ouellet and Sherk. Their formula is:

$$Y = -67.62 + 1.734X_1 + 0.1868X_2 + 69.77X_3 + 1.256X_4 + 0.006119X_5 + 22.37X_6 - 0.01832X_7$$

where: Y = estimated index of suitability X_1 = monthly mean of the daily minimum temperatures (°C) of the coldest month X_2 = mean frost free period above 0 °C in days X_3 = amount of rainfall (R) from June to November, inclusive, in terms of $R/(R+a)$ where $a=25.4$ if R is in millimeters and $a=1$ if R is in inches X_4 = monthly mean of the daily maximum temperatures (°C) of the warmest month X_5 = winter factor expressed in terms of $(0\ °C - X_1)R_{jan}$ where Rjan represents the rainfall in January expressed in mm X_6 = mean maximum snow depth in terms of $S/(S+a)$ where $a=25.4$ if S is in millimeters and $a=1$ if S is in inches X_7 = maximum wind gust in (km/hr) in 30 years

For practical purposes, Canada has adopted the American hardiness zone classification system.

American Horticultural Society Heat Zones

In addition to the USDA Hardiness zones there are the American Horticultural Society (AHS) heat zones.

The criterion is the average number of days per year when the temperature exceeds 30 °C (86 °F). The AHS Heat Zone Map for the US is available on the American Horticultural Society website.

Zone	From	To
1	< 1	
2	1	7
3	8	14
4	15	30
5	31	45
6	46	60
7	61	90
8	91	120
9	121	150
10	151	180
11	181	210
12	>210	

European Cities (AHS heat zones)

City	Zone	City	Zone
Amsterdam, The Netherlands	**2**	Antwerp, Belgium	**2**
Belfast, Northern Ireland	**1**	Berlin, Germany	**3**
Bratislava, Slovakia	**4**	Bucharest, Romania	**6**
Cardiff, Wales	**1**	Catania, Italy	**8**
Belgrade **6**			
Copenhagen, Denmark	**2**	Cork, Ireland	**1**
Derry, Northern Ireland	**1**	Dublin, Ireland	**1**
Düsseldorf, Germany	**3**	Edinburgh, Scotland	**1**
Gdańsk, Poland	**2**	Galway, Ireland	**1**
Glasgow, Scotland	**1**	Hamburg, Germany	**2**
Helsinki, Finland	**2**	Istanbul, Turkey	**6**
Kaliningrad, Russia	**2**	Kiev, Ukraine	**4**
Kraków, Poland	**4**	Lisbon, Portugal	**7**
Ljubljana, Slovenia	**6**	London, England	**2**

Madrid, Spain	7	Málaga, Spain	7
Marseille, France	7	Milan, Italy	6
Minsk, Belarus	3	Moscow, Russia	2
Munich, Germany	3	Murmansk, Russia	1
Nicosia, Cyprus	9	Oslo, Norway	2
Oulu, Finland	1	Palma, Spain	8
Paris, France	3	Perm, Russia	3
Prague, Czech Republic	3	Reykjavík, Iceland	1
Riga, Latvia	2	Rome, Italy	7
Rovaniemi, Finland	1	Saint Petersburg, Russia	2
Sarajevo, Bosnia and Herzegovina	5	Santander, Spain	3
Simrishamn, Sweden	1	Sochi, Russia	6
Sofia, Bulgaria	6	Stockholm, Sweden	2
Strasbourg, France	4	Tallinn, Estonia	2
Tuapse, Russia	7	Tórshavn, Faroe Islands	1
Tromsø, Norway	1	Trondheim, Norway	1
Umeå, Sweden	1	Vienna, Austria	4
Vilnius, Lithuania	2	Vorkuta, Russia	1
Warsaw, Poland	3	Zürich, Switzerland	4

MICROCLIMATE

A microclimate is a local set of atmospheric conditions that differ from those in the surrounding areas. Because climate is a statistic, there is a variation with the mean values of spatial and temporal parameters. Thus, even within a region, there are sets of statistically distinct conditions occurring through time (i.e. microclimates). The term may refer to areas as small as a few square meters or square feet (for example a garden bed) or area as large as many square kilometers or square miles.

Microclimates can be found in most places. They may exist near bodies of water which may cool the local atmosphere. They may also exist in heavy urban areas where brick, concrete, and asphalt absorb the sun's energy, heat up, and re-radiate the absorbed heat the ambient air. This results to an urban heat island, which is an example of a microclimate. Urban heat islands may also occur in places where the ground is made with tar or concrete. Man-made objects like tar and concrete do not take in much heat, and instead, radiate it back to the environment, ultimately forming an urban heat island.

Another contributing factor of microclimate is the slope or aspect of an area. South-facing slopes in the Northern Hemisphere and north-facing slopes in the Southern Hemisphere are exposed to more direct sunlight than their opposite slopes, and are therefore warmer for longer periods of time, giving the slope a warmer microclimate than the areas around the slope.

Microclimate on rock located in intertidal zone in Sunrise-on-Sea, South Africa

Background

Tree ferns thrive in a protected dell area in the Lost Gardens of Heligan, in Cornwall, England, latitude 50° 15'N.

The terminology "micro-climate" first appeared in the 1950s in publications such as *Climates in Miniature: A Study of Micro-Climate Environment* (Thomas Bedford Franklin, 1955).

The area in a developed industrial park may vary greatly from a wooded park nearby, as natural flora in parks absorb light and heat in leaves that a building roof or parking lot just radiates back into the air. Advocates of solar energy argue that widespread use of solar collection can mitigate overheating of urban environments by absorbing sunlight and putting it to work instead of heating the foreign surface objects.

A microclimate can offer an opportunity as a small growing region for crops that cannot thrive in the broader area; this concept is often used in permaculture practiced in northern temperate climates. Microclimates can be used to the advantage of gardeners who carefully choose and position their plants. Cities often raise the average temperature by zoning, and a sheltered position can reduce the severity of winter. Roof gardening, however, exposes plants to more extreme temperatures in both summer and winter.

Tall buildings create their own microclimate, both by overshadowing large areas and by channeling strong winds to ground level. Wind effects around tall buildings are assessed as part of a microclimate study.

Microclimates can also refer to purpose-made environments, such as those in a room or other enclosure. Microclimates are commonly created and carefully maintained in museum display and storage environments. This can be done using passive methods, such as silica gel, or with active microclimate control devices.

Usually, if the inland areas have a humid continental climate, the coastal areas stay much milder during winter months, in contrast to the hotter summers. This is the case further north on the American west coast, such as in British Columbia, Canada, where Vancouver has an oceanic wet winter with rare frosts, but inland areas that average several degrees warmer in summer have cold and snowy winters.

Soil Types

The type of soil found in an area can also affect microclimates. For example, soils heavy in clay can act like pavement, moderating the near ground temperature. On the other hand, if soil has many air pockets, then the heat could be trapped underneath the topsoil, resulting in the increased possibility of frost at ground level.

Sources and Influences On Microclimate

Two main parameters to define a microclimate within a certain area are temperature and humidity. A source of a drop in temperature and/or humidity can be attributed to different sources or influences. Often microclimate is shaped by a conglomerate of different influences and is a subject of microscale meteorology.

Cold Air Pool

The well known examples of cold air pool (CAP) effect are Gstettneralm Sinkhole in Austria (lowest recorded temperature -53 C) and Peter Sinks in the US. The main criterion on the wind speed v in order to create a warm air flow penetration into a CAP is the following:

$$\mathrm{Fr} = \frac{v}{Nh} \geq \mathrm{Fr}_c,$$

where Fr is the Froude number, N --- the Brunt–Väisälä frequency, h --- depth of the valley, and Fr_c --- Froude number at the threshold wind speed.

Craters

The presence of permafrost close to the surface in a crater creates a unique microclimate environment.

Caves and Lava Tubes

Lava tubes, whiich formed due to volcanic activity, and caves have similarity with each other. However, the microclimate in lava caves has a more dominant presence of basalt. Lava tubes and basaltic caves are important astrobiological targets on Earth and Mars (see also Martian lava tube).

Plant Microclimate

As pointed out by Rudolf Geiger in his book , while climate influences the living plant, the opposite effect of plants affecting their environment can also take place. This is known as plant climate.

Dams

Artificial reservoirs as well as natural ones create microclimates and often influence the macroscopic climate as well.

Cities and Regions Known for Microclimates

Americas

- Northern California above the Bay Area is also well known for microclimates with significant differences of temperatures. The coastline typically averages between 17 °C (63 °F) and 19 °C (66 °F) during summer months along that coastline, but inland towns not far from the ocean such as Lakeport, average as much as 34 °C (93 °F) in spite of being just around 40 miles inland. Even as far north as the Klamath River valley around the 41st parallel north between Willow Creek and Eureka averages such temperatures, which is extremely hot for such northerly areas. At this parallel, the temperature at the coast is so cool that Willow Creek beats Eureka's all-time record temperature on average 79 times per year. This is in spite of the areas being less than 50 miles from each other. In oceanic stable terms it is like traveling from the summers in the north of England to the south of Spain in a fraction of the distance.

- San Francisco is a city with microclimates and submicroclimates. Due to the city's varied topography and influence from the prevailing summer marine layer, weather conditions can vary by as much as 9 °F (5 °C) from block to block. The Noe Valley district for example, is typically warmer and sunnier than adjacent areas because the surrounding hills block some of the cool fog from the Pacific.
- The region as a whole, known as the San Francisco Bay area can have a wide range of extremes in temperature. In the basins and valleys adjoining the coast, climate is subject to wide variations within short distances as a result of the influence of topography on the circulation of marine air. The San Francisco Bay Area offers many varieties of climate within a few miles. In the Bay area, for example, the average maximum temperature in July is about 64 °F (18 °C) at Half Moon Bay on the coast, 87 °F (31 °C) at Walnut Creek only 25 miles (40 km) inland, and 95 °F (35 °C) at Tracy, just 50 miles (80 km) inland.

- The Los Angeles and San Diego areas are also subject to phenomena typical of a microclimate. The temperatures can vary as much as 18 °F (10 °C) between inland areas and the coast, with a temperature gradient of over one degree per mile (1.6 km) from the coast inland. Hills and mountains can also block coastal air masses. The San Fernando Valley is usually much warmer in summer than most of Los Angeles, because the Santa Monica Mountains usually block the cool ocean breezes and fog. Southern California has also a weather phenomenon called "June Gloom" or "May Grey", which sometimes gives overcast or foggy skies in the morning at the coast, but usually gives sunny skies by noon, during late spring and early summer.
- Calgary, Alberta is also known for its microclimates. Especially notable are the differences between the downtown and river valley/flood plain regions and the areas to the west and north. This is largely due to an elevation difference within the city's boundaries of over 1,000 ft (300 m), but can also be attributed somewhat, to the effects of the seasonal Chinooks.
- Halifax, Nova Scotia also has numerous microclimates. Coastal temperatures and weather conditions can differ considerably from areas located just 5–15 km inland. This is true in all seasons. Varying elevations are common throughout the city, and it is even possible to experience several microclimates while traveling on a single highway due to these changing elevations.
- Chesapeake Bay is also known for its subtropical microclimate. It is most notable for its mild climatic effects on the area east and west of the lowlands of Maryland and Delmarva. Having over 64,000 square miles of water; (most of which is a mix of fresh and salt water) creates higher levels of humidity and heat in the spring and summer months. An example of this effect is the survival of tropical palm trees and plants such as water hyacinths in the area.
- Santiago, Chile and Villa de Merlo, Argentina are also subject to microclimates.

Europe

- Known for its wines, the Ticino region in Switzerland benefits from a microclimate in which palm trees and banana trees grow.
- Gran Canaria is called "Miniature Continent" for its rich variety of microclimates.
- Biddulph Grange is very rich with microclimates as a result of the large dips and variety of very large trees alongside a large amount of water.
- Leeds, located in England is known to have a number of microclimates because of the number of valleys surrounding the city centre.
- The coastal areas in the Andalusia region of Spain typically average around at 30 °C (86 °F) in summer, but Tarifa only averages 24 °C (75 °F). Further north along the coast Cádiz has a summer average of 27 °C (81 °F) with very warm nights, whereas nearby Jerez de la Frontera has summer highs of 33 °C (91 °F) with inland areas further north such as Seville being even hotter.

Asia and Oceania

- Amman, Jordan has extreme examples of microclimate, and almost every neighbourhood exhibits its own weather. It is known among locals that some boroughs such as the northern and western suburbs are among the coldest in the city, and can be experiencing frost or snow whilst other warmer districts such as the city centre can be at much warmer temperatures at the same time.
- Yazd and Kashan in Iran, the traditional architecture benefits from central courtyards with trees and water pools that function together with wind catchers to create a favorable microclimate in this desert area.
- Sydney, Australia has a microclimate occurring prominently in the warmer months. Inland, in Sydney's western suburbs, the climate is drier and significantly hotter with temperatures generally around 3–7 °C (5–13 °F) above Sydney CBD and Eastern Suburbs (the coast), as sea breezes don't penetrate further inland. In summer, the coast averages at 25.9 °C (78.6 °F), while inland varies between 28 °C (82 °F) to 30 °C (86 °F), depending on the suburb. In extreme occasions, the Coast would have a temperature of 24 °C (75 °F), while a suburb 20 km (19 mi) inland bakes in 36 °C (97 °F) heat. However, winter lows in the West are around 3–5 °C (5–9 °F) cooler than the coastal suburbs, and may provide mild to moderate frost. Within the city and surrounds, rainfall varies, from around 682.5 mm (26.87 in) in the far-west to 1,213.8 mm (47.79 in) at Observatory Hill (the east, or the coast).

REFERENCES

1. Titchmarsh, Alan. "How to be a gardener". BBC. Retrieved 24 April 2013.
2. "How Woody Plants Survive Extreme Cold". Iowa State University. Horticulture & Home Pest News. March 1, 1996. Retrieved December 4, 2014.
3. http://www.rhs.org.uk/Media/PDFs/News/2012_RHS-Hardiness-Rating

4. http://www.metoffice.gov.uk/climate/uk/averages/ukmapavge.html
5. Thorsen, Stig Morten; Höglind, Mats (2010-12-15). "Modelling cold hardening and dehardening in timothy. Sensitivity analysis and Bayesian model comparison". Agricultural and Forest Meteorology. 150 (12): 1529–1542. doi:10.1016/j.agrformet.2010.08.001.
6. Smallwood, Maggie; Bowles, Dianna J. (2002-07-29). "Plants in a cold climate". Philosophical Transactions of the Royal Society B: Biological Sciences. 357 (1423): 831–847. doi:10.1098/rstb.2002.1073. ISSN 0962-8436. PMC 1692998 . PMID 12171647.
7. McKhann, Heather I.; Gery, Carine; Bérard, Aurélie; Lévêque, Sylvie; Zuther, Ellen; Hincha, Dirk K.; De Mita, S.; Brunel, Dominique; Téoulé, Evelyne (2008-01-01). "Natural variation in CBF gene sequence, gene expression and freezing tolerance in the Versailles core collection of Arabidopsis thaliana". BMC Plant Biology. 8: 105. doi:10.1186/1471-2229-8-105. ISSN 1471-2229. PMC 2579297 . PMID 18922165.
8. Forbes, James C.; Watson, Drennan (1992-08-20). Plants in Agriculture. Cambridge University Press. ISBN 9780521427913.
9. Chen, CP; Denlinger, DL; Lee, RE. "A rapid cold-hardening process in insects.". Science. 238: 1415–7. doi:10.1126/science.238.4832.1415. PMID 17800568.
10. Lee, RE; Czajka, MC. "A rapid cold-hardening response protecting against cold shock injury in Drosophila melanogaster.". J Exp Biol. Science. 148: 245–54. PMID 2106564.
11. Duman, J (2002). "The inhibition of ice nucleators by insect antifreeze proteins is enhanced by glycerol and citrate". Journal of Comparative Physiology B. 172: 163–168. doi:10.1007/s00360-001-0239-7.
12. Overgaard, J.; Sørensen, J. G.; Com, E.; Colinet, H. (2013). "The rapid cold hardening response of Drosophila melanogaster: Complex regulation across different levels of biological organization". Journal of Insect Physiology. 62: 46–53. doi:10.1016/j.jinsphys.2014.01.009.
13. Lee, R. E.; Damodaran, K.; Yi, S. X.; Lorigan, G. A. (2006). "Rapid Cold-Hardening Increases Membrane Fluidity and Cold Tolerance of Insect Cells". Cryobiology. 52 (3): 459–463. doi:10.1016/j.cryobiol.2006.03.003.
14. Park, Y.; Kim, Y. (2014). "A specific glycerol kinase induces rapid cold hardening of the diamondback moth, Plutella xylostella". Journal of Insect Physiology. 67: 56–63. doi:10.1016/j.jinsphys.2014.06.010
15. Sunset National Garden Book. Sunset Books Inc. Menlo Park, California (1997)
16. "RHS hardiness rating". RHS hardiness rating. Royal Horticultural Society. Retrieved 24 September 2016.
17. http://subidaimagen.infojardin.com/subire/images/opt1366025961n.JPG
18. "Hardiness zone - Gardenology.org - Plant Encyclopedia and Gardening wiki". Gardenology.org. Retrieved 2012-10-13.

19. "Europa Hardiness zone map". Backyardgardener.com. Retrieved 2012-10-13.
20. "Hardiness Zones". Havlis.cz. Retrieved 2012-10-13.
21. "New arborday.org Hardiness Zone Map reflects warmer climate : Latest hardiness zones, based on most current temperature data available, suggest up-to-date choices for best trees to plant". Retrieved 2007-12-27.
22. "USDA Plant Hardiness Zone Map". USDA Agricultural Research Service.
23. "Plant Hardiness Zones for Australia". Retrieved 2010-11-11.
24. "Canada's Plant Hardiness Site". Retrieved 2015-10-01.
25. Ouellet, C.E., Sherk, L.C. 1967a. Woody ornamental plant zonation I. Indices of winter hardiness. Can J. Plant Sci. 47:231-238.
26. Ouellet, C.E., Sherk, L.C. 1967b. Woody ornamental plant zonation. II. Suitability indices of localities. Can J. Plant Sci. 47: 339-349.
27. Ouellet, C.E., Sherk, L.C. 1967c. Woody ornamental plant zonation III. Suitability map for the probable winter survival of ornamental trees and shrubs. Can J. Plant Sci. 47: 351-358.
28. "AHS Plant Heat Zone Map". GardeningPlaces. Retrieved 2012-10-13.
29. "American Horticultural Society: Plant Heat-Zone Map"(PDF). ahs.org. Retrieved 2012-10-13.
30. PlantZAfrica 2016, Horticultural Zones
31. http://www.amazon.co.uk/Thomas-Bedford-Franklin/e/B00JWR3UW2
32. Gardening Resources, Cornell University
33. Microclimate, www.wetter-freizeit.com, in German.
34. J. Racovec et al. Turbulent dissipation of the cold-air pool in a basin: comparison of observed and simulated development. Meteorol. Atmos. Phys. 79, 195-213 (2002).
35. Permafrost in Hawaii, NASA Astrobiology Institute, 2010
36. R. Geiger. The climate near the ground. Harvard University Press, 1957.
37. "Climate Of California". Wrcc.dri.edu. Western Regional Climate Center. Retrieved 2014-02-02.
38. "Calgary's Gardening Climate". Scarboro.ca. 2008-05-05. Retrieved 2014-02-02.
39. http://extension.umd.edu/learn/august-ornamental-plant-tips
40. http://www.cbf.org/about-the-bay/more-than-just-the-bay
41. http://pdxscholar.library.pdx.edu/iss_pub/46/
42. Potter, Robert B.; Khadija Darmame; Nasim Barham; Stephen Nortcliff (2008). "Ever-growing Amman, Jordan: Urban expansion, social polarisation and contemporary urban planning issues" (PDF). Habitat International. www.journals.elsevier.com/habitat-international. 33: 81–92. doi:10.1016/j.habitatint.2008.05.005. Retrieved 2014-02-02.

43. "Sydney/Kingsford-Smith International Airport". Climate statistics for Australian locations. Bureau of Meteorology. Retrieved 27 August 2014.
44. "Penrith". Climate statistics for Australian locations. Bureau of Meteorology. Retrieved 19 January 2014.
45. Sydney's Climate
46. "Badgerys Creek AWS". Climate statistics for Australian locations. Bureau of Meteorology. Retrieved 19 January 2014.

CHAPTER 5

Desiccation Tolerance

Chapter Contents

- **Introduction**
- **Desiccation Tolerance**
- **Crop Desiccation**
- **Glyphosate**

Desiccation tolerance is found throughout the bacteria, fungi, animal and plant kingdoms [1,2]. In the plant kingdom, the organisms that commonly display tolerance to dessication are non-tracheophytes, such as mosses, and seeds. As opposed to drought tolerance, which involves surviving moderate water loss (e.g. 90% relative water content [RWC]), desiccation tolerance, is the ability to survive absolute water contents of 0.1 g H_2O g1.

INTRODUCTION

Desiccation is the state of extreme dryness, or the process of extreme drying. A desiccant is a hygroscopic (attracts and holds water) substance that induces or sustains such a state in its local vicinity in a moderately sealed container.

Desiccation cracks in sludge

Centripetal desiccation cracks in the Lower Jurassic Moenave Formation at the St. George Dinosaur Discovery Site at Johnson Farm, southwestern Utah. A dinosaur footprint is at center.

Chemistry

A desiccator is a heavy glass or plastic container used in practical chemistry for making or keeping small amounts of materials very dry. The material is placed on a shelf, and a drying agent or *desiccant*, such as dry silica gel or anhydrous sodium hydroxide, is placed below the shelf.

Often some sort of humidity indicator is included in the desiccator to show, by color changes, the level of humidity. These indicators are in the form of indicator plugs or indicator cards. The active chemical is cobalt chloride ($CoCl_2$). Anhydrous cobalt chloride is blue. When it bonds with two water molecules, ($CoCl_2{\bullet}2H_2O$), it turns purple. Further hydration results in the pink hexaaquacobalt(II) chloride complex $[Co(H_2O)_6]^{2+}$.

A vacuum desiccator (left) and desiccator (right). Silica gel with cobalt chloride indicator placed in the lower shelf is used as the desiccant.

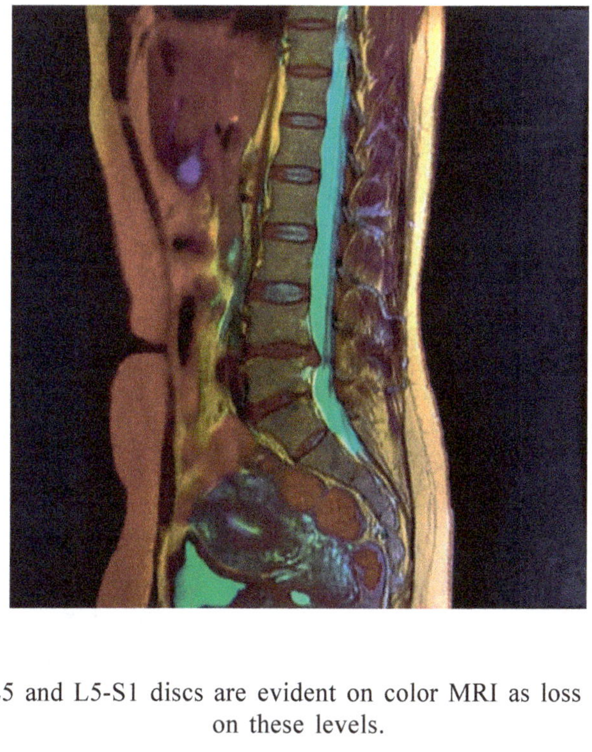

Desiccation of L4-L5 and L5-S1 discs are evident on color MRI as loss of blue color visible on these levels.

Biology and Ecology

In biology and ecology, desiccation refers to the drying out of a living organism, such as when aquatic animals are taken out of water, slugs are exposed to salt, or when plants are exposed to sunlight or drought. Ecologists frequently study and assess various organisms' susceptibility to desiccation. For example, in the study performed by Erkut et al., the investigators found that Caenorhabditis elegans dauer is a true anhydrobiote that can withstand extreme desiccation and that the basis of this ability is founded in the metabolism of trehalose.

DESICCATION TOLERANCE

Desiccation tolerance refers to the ability of an organism to withstand or endure extreme dryness, or drought-like conditions. Plants and animals living in arid or periodically arid environments such as temporary streams or ponds may face the challenge of desiccation, therefore physiological or behavioral adaptations to withstand these periods are necessary to ensure survival. In particular, insects occupy a wide range of ecologically diverse niches and, so, exhibit a variety of strategies to avoid desiccation.

In general, desiccation resistance in insects is measured by the change in mass during dry conditions. The overall mass difference between measurements before and after aridity exposure is attributed to body water loss, as respiratory water loss is generally considered negligible.

Types of Desiccation Resistance

There are three main ways in which insects can increase their tolerance to desiccation: by increasing their total body water content; by reducing the rate of body water loss; and by tolerating a larger proportion of overall water loss from the body.Survival time is determined by initial water content, and can be calculated by dividing water loss tolerance (the maximum amount of water that may be removed without resulting in death) by water loss rate.

Increasing Body Water Content

Insects with a higher initial body water content have better survival rates during arid conditions than insects with a lower initial body water content. Higher amounts of internal body water lengthen the time necessary to remove the amount of water required to kill the organism. The way in which body water content is increased may differ depending on the species.

The accumulation of glycogen during the insect larval stage has been linked to increased body water content and is likely a source of metabolic water during dry conditions. Glycogen, a glucose polysaccharide, acts as an oxidative energy source during times of physiological stress. Because it binds up to five times its weight in bulk water, insects with increased levels of body glycogen also have higher amounts of internal water. In general, insects selected for desiccation resistance also exhibit longer larval stages than those sensitive to desiccation. This increase in development time is likely a response to the environment, allowing larvae more time to accumulate glycogen, and therefore more water before eclosion.

Another possible source contributing to higher levels of initial body water in insects is hemolymph volume. The insect equivalent to blood, hemolymph is the fluid found within the hemocoel, and is the largest pool of extracellular water within the insect body. In the fruit-fly *Drosophila melanogaster*, flies selected for desiccation resistance also yielded higher amounts of hemolymph. Higher hemolymph volume is linked to an increase in carbohydrates, in particular trehalose, a common sugar found in many plants and animals with high desiccation resistance. *Drosophila melanogaster* flies selected for desiccation resistance show a 300% increase in hemolymph volume compared to control flies, correlating to a similar increase in trehalose levels. During periods of aridity, cells dehydrate and draw upon hemolymph stores to replenish intracellular water; therefore, insects with higher levels of this fluid are less prone to desiccation.

Insects may also increase body water content by simply feeding more often. Because sugar is slowly absorbed into the hemolymph at each meal, increasing the frequency at which the insect ingests a sugar source also increases its desiccation tolerance. Furthermore, the crop may also act not only to store food prior to digestion but to provide an additional reservoir for water and sugar.

Reducing Rate of Water Loss

Another strategy used to reduce the risk of death by dehydration is to reduce the rate at which water is lost. The three main ways through which insects can lose water are (1) the surface of the body (integument); (2) the tracheae (respiration); and (3) excretion, or waste products. The important feature in reducing water loss in land snails during inactivity is an epiphragm.

Integument

The exoskeleton or integument of insects acts as an impermeable, protective layer against desiccation. It is composed of an outer epicuticle, underlain by an procuticle that itself may be further divided into an exo- and endocuticle. The endocuticle provides the insect with toughness and flexibility and the hard exocuticle serves to protect vulnerable body parts. However, the outer cuticular layer (epicuticle) is a protein-polyphenol complex made up of lipoproteins, fatty acids, and waxy molecules, and is the insect's primary defense against water loss. Many insect orders secrete an additional cement layer over their wax layer, likely to protect against the abrasion or removal of waxy molecules. This layer is composed of lipids and proteins held together by polyphenolic compounds and is secreted by the dermal glands.

In general, the rate of water loss in insects is low at moderate temperatures. Once a species-specific critical temperature (Tc) is reached, as temperatures continue to increase, rapid water loss occurs. The "lipid melting model" is used to explain this sudden increase in the rate of water loss. The lipid melting model states that increased cuticular water loss is directly related to the melting of surface lipids. Insects already adapted to more arid environments have a higher Tc; that is, their cuticular properties change and lipid structures melt at a higher critical temperature.

In some insects, the rate of cuticular water loss is controlled to some extent by the neuroendocrine system. Immediately following head removal, decapitated cockroaches exhibit a large increase in transpiration across the cuticle, leading to severe dehydration. Injection of brain hormones into freshly separated bodies results in a sharp reduction in cuticular water loss.

Tracheae

In general, insects adapted to arid environments also have an impermeable cuticular membrane that prevents water loss. Therefore, a majority of water lost to the atmosphere occurs via the air-filled tracheae. To help reduce water loss, many insects have outer coverings to their tracheae, or spiracles, which shut when open respiration is unnecessary and prevent water from escaping. Insects at a greater risk for water loss face the challenge of either a depleted oxygen supply or desiccation, leading to an adaptive increase in tracheal volume in order to receive more oxygen.

Excretion

Following feeding, most insects retain enough water to completely hydrate their bodies, excreting the remainder. However, the amount of water excreted differs between species, and depends on the relative humidity and dryness of the environment. For example, Tsetse flies maintained at a high relative humidity, and thus non-arid conditions, excrete fecal matter with approximately 75% water content, whereas Tsetse flies maintained at a low relative humidity, and thus dry conditions, excrete fecal matter with only 35% water content. This adaptation helps minimize water loss in unfavorable conditions and increase chances of survival.

Tolerating Greater Water Loss

Most insects can tolerate a 30-50% loss of body water; however, insects adapted to dry environments can tolerate a 40-60% loss of body water. Initial body size also plays a large role in how much water loss can be tolerated, and, in general, larger insects can tolerate a larger percentage of body water loss than smaller insects. The female beetle *Alphitobius diaperinus*, for example, is larger than its male counterpart and can thus tolerate 4% more water loss. It is hypothesized that larger insects have increased lipid reserves, preventing dehydration and desiccation.

Behavior Modification

In addition to physiological adaptations that increase desiccation resistance, behavioral responses of insects to arid environments significantly decrease dehydration potential. *Drosophila melanogaster* fruit flies, for example, will actively move to areas with higher atmospheric water content when placed in dry environments. Also, the dung beetle buries food in underground chambers, thereby ensuring water and energy sources during periodically dry conditions. Feeding location may also be altered to ensure body hydration. Some caterpillars preferentially feed on the underside of leaves, where microclimate has higher relative humidity. In a highly time-consuming activity such as feeding, these insects significantly reduce their chances of desiccation.

Cryptobiosis (Anhydrobiosis)

Cryptobiosis refers to the state of an organism that has no detectable metabolic activity, resulting from extreme and unfavorable environmental conditions; anhydrobiosis refers to the state of surviving the loss of (almost) all body water. Although this state is commonly observed in invertebrates, only one insect is known to be cryptobiotic (anhydrobiotic), the African chironomid *Polypedilum vanderplanki*. *Polypedilum vanderplanki* undergoes anhydrobiosis, a cryptobiotic state wherein the body is completely dehydrated. The larvae of *P. vanderplanki* inhabit rock pools that commonly dry out completely. In response, *P. vanderplanki* larvae enter an anhydrobiotic state, during which changes in body osmolarity trigger the production of large amounts of trehalose. Due to its capacity for water replacement and vitrification, the accumulation of trehalose prevents the death of the larvae from water loss.

DNA Damage and Repair

Several bacterial species have been shown to accumulate DNA damages upon desiccation. *Deinococcus radiodurans* is extremely resistant to ionizing radiation. The functions necessary to survive ionizing radiation are also necessary to survive prolonged desiccation. Radiation resistance is considered to be an incidental consequence of the organism's evolutionary adaptation to dehydration, a common physiological stress in nature. The chromosomal DNA from desiccated *D. radiodurans* revealed increased DNA double-strand breaks. DNA double-strand breaks are repaired principally by a RecA-dependent recombination process that requires the presence of two genome copies. By this process *D. radiodurans* can survive thousands of double-strand breaks per cell.

Mycobacterium smegmatis mutant strains that are deficient in the ability to repair double-strand breaks by the non-homologous enjoining (NHEJ) pathway are more sensitive to prolonged desiccation during stationary phase than wild-type strains. NHEJ appears to be the preferred pathway for repairing double-strand breaks caused by desiccation during stationary phase. NHEJ can repair double-strand breaks even when only one chromosome is present in a cell.

Upon exposure to extreme dryness, *Bacillus subtilis* spores acquire DNA-double strand breaks and DNA-protein crosslinks.

Broadcasting

In broadcast engineering, a desiccator may be used to pressurize the feedline of a high-power transmitter. Because it carries a large amount of energy from the transmitter to the antenna, the feedline must have low dielectric losses. Because it must also be lightweight so as not to overload the radio tower, air is often used as the dielectric. Since moisture can condense in these lines, desiccated air or nitrogen gas is pumped in. This pressure also keeps water or other dampness from coming in the line at any point along its length.

CROP DESICCATION

Pre-harvest **crop desiccation** (also **siccation**) refers to the application of a herbicide to a crop shortly before harvest. Herbicides used include glyphosate, diquat and glufosinate. For potatoes, carfentrazone-ethyl is used. Other desiccants are cyanamide, cinidon-ethyl, and pyraflufen-ethyl. Uneven crop growth is a problem in northern climates during wet summers or when weed control is poor and several of advantages of desiccation have been cited: More even ripening is achieved and harvest can be conducted earlier; weed control is initiated for a future crop; earlier ripening allows for earlier replanting; desiccation reduces green material in the harvest putting less strain on harvesting machinery. Some crop may be mechanically destroyed when crop desiccation machinery moves through the field.

The application of glyphosate differs between countries significantly. It is commonly used in the UK where summers are wet and crops may ripen unevenly. Thus in the UK 78% of oilseed rape is desiccated before harvest, but only 4% in Germany.

Desiccated potato plants prior to harvest

Use

Pre-harvest desiccation is used in agriculture in crops including:

- Maize
- Cereals such as barley, wheat, and oats
- Oilseed rape
- Legumes including lentils, garbanzos, and soybeans
- Sunflower
- Potatoes[:474]
- Cotton[:24–26]

GLYPHOSATE

Glyphosate (*N*-(phosphonomethyl)glycine) is a broad-spectrum systemic herbicide and crop desiccant. It is an organophosphorus compound, specifically a phosphonate. It is used to kill weeds, especially annual broadleaf weeds and grasses that compete with crops. It was discovered to be an herbicide by Monsanto chemist John E. Franz in 1970. Monsanto brought it to market in 1974 under the trade name **Roundup**, and Monsanto's last commercially relevant United States patent expired in 2000.

Farmers quickly adopted glyphosate, especially after Monsanto introduced glyphosate-resistant Roundup Ready crops, enabling farmers to kill weeds without killing their crops. In 2007, glyphosate was the most used herbicide in the United States' agricultural sector and the second-most used in home and garden, government and industry, and commerce.By 2016 there was a 100-fold increase from the late 1970s in the frequency of applications and volumes of glyphosate-based herbicides (GBHs) applied, partly in response to the unprecedented global emergence and spread of glyphosate-resistant weeds.[:1]

Glyphosate is absorbed through foliage, and minimally through roots, and transported to growing points. It inhibits a plant enzyme involved in the synthesis of three aromatic amino acids: tyrosine, tryptophan, and phenylalanine. Therefore, it is effective only on actively growing plants and is not effective as a pre-emergence herbicide. An increasing number of crops have been genetically engineered to be tolerant of glyphosate (e.g. Roundup Ready soybean, the first Roundup Ready crop, also created by Monsanto) which allows farmers to use glyphosate as a postemergence herbicide against weeds. The development of glyphosate resistance in weed species is emerging as a costly problem. While glyphosate and formulations such as Roundup have been approved by regulatory bodies worldwide, concerns about their effects on humans and the environment persist.

Many regulatory and scholarly reviews have evaluated the relative toxicity of glyphosate as an herbicide. The German Federal Institute for Risk Assessment toxicology review in 2013 found that "the available data is contradictory and far from being convincing" with regard to correlations between exposure to glyphosate formulations and risk of various cancers, including non-Hodgkin lymphoma (NHL). A meta-analysis published in 2014 identified an increased risk of NHL in workers exposed to glyphosate formulations. In March 2015 the World Health Organization's International Agency for Research on

Cancer classified glyphosate as "probably carcinogenic in humans" (category 2A) based on epidemiological studies, animal studies, and *in vitro* studies. In November, 2015, the European Food Safety Authority published an updated assessment report on glyphosate, concluding that "the substance is unlikely to be genotoxic (i.e. damaging to DNA) or to pose a carcinogenic threat to humans." Furthermore, the final report clarified that while other, probably carcinogenic, glyphosate-containing formulations may exist, studies "that look solely at the active substance glyphosate do not show this effect." In May 2016, the Joint FAO/WHO Meeting on Pesticide Residues concluded that "glyphosate is unlikely to pose a carcinogenic risk to humans from exposure through the diet", even at doses as high as 2,000 mg/kg body weight orally.

Discovery

Glyphosate was first synthesized in 1950 by Swiss chemist Henry Martin, who worked for the Swiss company Cilag. The work was never published.[1] Stauffer Chemical patented the agent as a chemical chelator in 1964 as it binds and removes minerals such as calcium, magnesium, manganese, copper, and zinc.

Somewhat later, glyphosate was independently discovered at Monsanto in 1970. Monsanto chemists had synthesized about 100 analogs of aminomethylphosphonic acid as potential water-softening agents. Two were found to have weak herbicidal activity, and John E. Franz, a chemist at Monsanto, was asked to try to make analogs with stronger herbicidal activity. Glyphosate was the third analog he made.[1–2] Franz received the National Medal of Technology in 1987 and the Perkin Medal for Applied Chemistry in 1990 for his discoveries. In 2008, USDA-ARS scientist Stephen O. Duke and Stephen B. Powles, an Australian weed expert — described glyphosate as a "virtually ideal" herbicide. In 2010 Powles stated: "glyphosate is a one in a 100-year discovery that is as important for reliable global food production as penicillin is for battling disease."

Chemistry

Glyphosate is an aminophosphonic analogue of the natural amino acid glycine, and like all amino acids, exists in different ionic states depending on pH. Both the phosphonic acid and carboxylic acid moieties can be ionised and the amine group can be protonated and the substance exists as a series of zwitterions. Glyphosate is soluble in water to 12 g/l at room temperature. The original synthetic approach to glyphosate involved the reaction of phosphorus trichloride with formaldehyde followed by hydrolysis to yield a phosphonate. Glycine is then reacted with this phosphonate to yield glyphosate, and its name is taken as a contraction of the compounds used in this synthesis - viz. **gly**cine and a **phosphonate**.

$PCl_3 + H_2CO \rightarrow Cl_2P(=O)\text{-}CH_2Cl$

$Cl_2P(=O)\text{-}CH_2Cl + 2\ H_2O \rightarrow (HO)_2P(=O)\text{-}CH_2Cl + 2\ HCl$

$(HO)_2P(=O)\text{-}CH_2Cl + H_2N\text{-}CH_2\text{-}COOH \rightarrow (HO)_2P(=O)\text{-}CH_2\text{-}NH\text{-}CH_2\text{-}COOH + HCl$

The main deactivation path for glyphosate is hydrolysis to aminomethylphosphonic acid.

Industrial Synthesis

Two main approaches are used to synthesize glyphosate industrially. The first is to react iminodiacetic acid with phosphorous acid and hydrochloric acid (sometimes formed *in situ* by addition of phosphorus trichloride) via a modified Mannich reaction. Oxidation then leads to the desired glyphosate product. Iminodiacetic acid is usually prepared on-site, such as by reaction of chloroacetic acid with ammonia and calcium hydroxide to produce the calcium iminodiacetate salt and then acidification of the product.

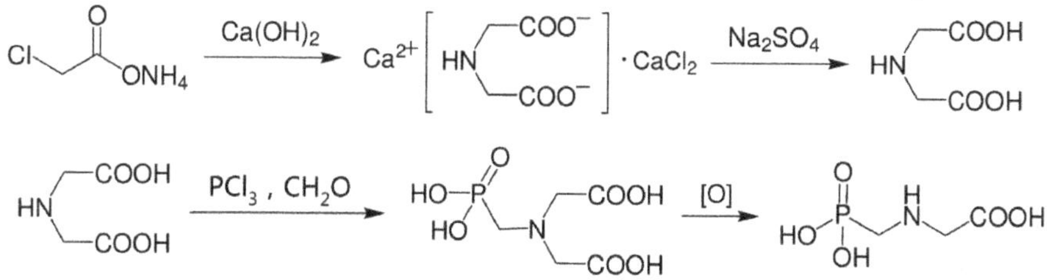

The chloroacetic acid approach is less efficient than other iminodiacetic acid approaches, owing to the production of calcium chloride waste and decreased yield. When hydrogen cyanide is readily available as a by-product (say), an alternative approach is to use iminodiacetonitrile, $HN(CH_2CN)_2$, and diethanolamine is also a suitable starting material.

The second involves the use of dimethyl phosphite in a one-pot synthesis. Glycine and paraformaldehyde are reacted in a suitable organic solvent (typically triethylamine and methanol) to produce bishydroxymethylglycine, $(HOCH_2)_2NCH_2COOH$. Dimethyl phosphite is then introduced and work-up with hydrochloric acid cleaves the hydroxymethyl group from the nitrogen atom whilst heating speeds the hydrolysis of both phosphate ester linkages.

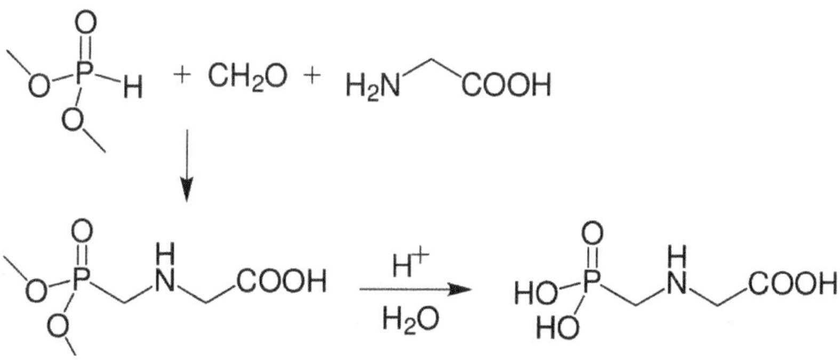

This synthetic approach is responsible for a substantial portion of the production of glyphosate in China, with considerable work having gone into recycling the triethylamine and methanol. Progress has also been made in attempting to eliminate the need for triethylamine altogether.

Mode of Action

Glyphosate kills plants by interfering with the synthesis of the aromatic amino acids phenylalanine, tyrosine, and tryptophan. It does this by inhibiting the enzyme 5-enolpyruvylshikimate-3-phosphate synthase (*EPSPS*), which catalyzes the reaction of shikimate-3-phosphate (S3P) and phosphoenolpyruvate to form 5-enolpyruvyl-shikimate-3-phosphate (EPSP). Glyphosate is absorbed through foliage and minimally through roots, meaning that it is only effective on actively growing plants and cannot prevent seeds from germinating. After application, glyphosate is readily transported around the plant to growing roots and leaves and this systemic activity is important for its effectiveness. Inhibiting the enzyme causes shikimate to accumulate in plant tissues and diverts energy and resources away from other processes. While growth stops within hours of application, it takes several days for the leaves to begin turning yellow.

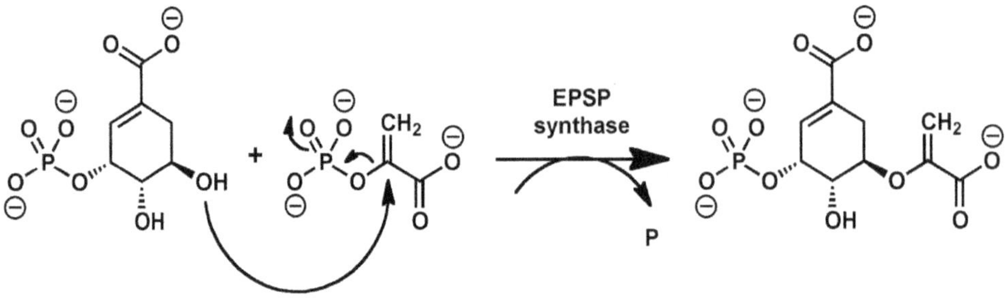

shikimate-3-phosphate phospho-enolpyruvate 5-enolpyruvylshikimate-3-phosphate

EPSP is subsequently dephosphorylated to chorismate, an essential precursor for the amino acids mentioned above. These amino acids are used in protein synthesis and to produce secondary metabolites such as folates, ubiquinones, and naphthoquinone.

X-ray crystallographic studies of glyphosate and EPSPS show that glyphosate functions by occupying the binding site of the phosphoenolpyruvate, mimicking an intermediate state of the ternary enzyme substrates complex. Glyphosate inhibits the EPSPS enzymes of different species of plants and microbes at different rates.

EPSPS is produced only by plants and microbes; the gene coding for it is not in the mammalian genome.

Environmental Fate

Glyphosate adsorbs strongly to soil, and residues are expected to generally be immobile in soil. Ground and surface water pollution is limited.Glyphosate is readily degraded by soil microbes to aminomethylphosphonic acid (AMPA, which like glyphosate strongly adsorbs to soil solids and is thus unlikely to leach to groundwater). Though both glyphosate and AMPA are commonly detected in water bodies, a portion of the AMPA detected may actually be the result of degradation of detergents rather than from glyphosate. Glyphosate does have the potential to contaminate surface waters due to its aquatic use patterns and through erosion, as it adsorbs to soil particles suspended in runoff. The mechanism of glyphosate sorption to soil is similar to that of phosphate fertilizers, the presence of which can reduce glyphosate sorption.Phosphate fertilizers are subject to release from sediments into water bodies under anaerobic conditions, and similar release can also occur with glyphosate, though significant impact of glyphosate release from sediments has not been established. Limited leaching can occur after high rainfall after application. If glyphosate reaches surface water, it is not broken down readily by water or sunlight.

The half-life of glyphosate in soil ranges between 2 and 197 days; a typical field half-life of 47 days has been suggested. Soil and climate conditions affect glyphosate's persistence in soil. The median half-life of glyphosate in water varies from a few to 91 days. At a site in Texas, half-life was as little as three days. A site in Iowa had a half-life of 141 days. The glyphosate metabolite AMPA has been found in Swedish forest soils up to two years after a glyphosate application. In this case, the persistence of AMPA was attributed to the soil being frozen for most of the year.Glyphosate adsorption to soil, and later release from soil, varies depending on the kind of soil. Glyphosate is generally less persistent in water than in soil, with 12- to 60-day persistence observed in Canadian ponds, although persistence of over a year has been recorded in the sediments of American ponds. The half-life of glyphosate in water is between 12 days and 10 weeks.

According to the National Pesticide Information Center fact sheet, glyphosate is not included in compounds tested for by the Food and Drug Administration's Pesticide Residue Monitoring Program, nor in the United States Department of Agriculture's Pesticide Data Program. However, a field test showed that lettuce, carrots, and barley contained glyphosate residues up to one year after the soil was treated with 3.71 lb of glyphosate per acre (4.15 kg per hectare). The U.S. has determined the acceptable daily intake of glyphosate at 1.75 milligrams per kilogram of bodyweight per day (mg/kg/bw/day) while the European Union has set it at 0.3.

Use

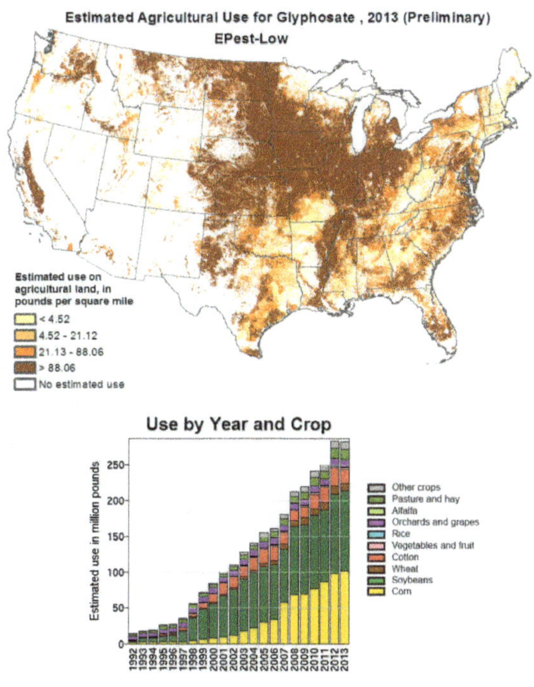

Estimated use of glyphosate in the USA in 2013 and estimated total use from 1992–2013

Glyphosate is effective in killing a wide variety of plants, including grasses and broadleaf and woody plants. By volume, it is one of the most widely used herbicides.In 2007, glyphosate was the most used herbicide in the United States agricultural sector, with 180 to 185 million pounds (82,000 to 84,000 tonnes) applied, the second-most used in home and garden with 5 to 8 million pounds (2,300 to 3,600 tonnes) and government applied 13 to 15 million pounds (5,900 to 6,800 tonnes) in industry and commerce. It is commonly used for agriculture, horticulture, viticulture, and silviculture purposes, as well as garden maintenance (including home use). It has a relatively small effect on some clover species and morning glory.

Glyphosate used as an alternative to mowing in an apple orchard in Ciardes, Italy

Glyphosate and related herbicides are often used in invasive species eradication and habitat restoration, especially to enhance native plant establishment in prairie ecosystems. The controlled application is usually combined with a selective herbicide and traditional methods of weed eradication such as mulching to achieve an optimal effect.

In many cities, glyphosate is sprayed along the sidewalks and streets, as well as crevices in between pavement where weeds often grow. However, up to 24% of glyphosate applied to hard surfaces can be run off by water. Glyphosate contamination of surface water is attributed to urban and agricultural use. Glyphosate is used to clear railroad tracks and get rid of unwanted aquatic vegetation. Since 1994, glyphosate has been used in aerial spraying in Colombia in coca eradication programs; Colombia announced in

May 2015 that by October, it would cease using glyphosate in these programs due to concerns about human toxicity of the chemical.

In addition to its use as an herbicide, glyphosate is also used for crop desiccation (siccation) to increase harvest yield, and as a result of desiccation, to increase sucrose concentration in sugarcane before harvest.

Genetically Modified Crops

Some micro-organisms have a version of 5-enolpyruvoyl-shikimate-3-phosphate synthetase (EPSPS) resistant to glyphosate inhibition. A version of the enzyme that was both resistant to glyphosate and that was still efficient enough to drive adequate plant growth was identified by Monsanto scientists after much trial and error in an *Agrobacterium* strain called CP4, which was found surviving in a waste-fed column at a glyphosate production facility.[56] This *CP4* EPSPS gene was cloned and transfected into soybeans. In 1996, genetically modified soybeans were made commercially available. Current glyphosate-resistant crops include soy, maize (corn), canola, alfalfa, sugar beets, and cotton, with wheat still under development.

In 2015, 89% of corn, 94% of soybeans, and 89% of cotton produced in the US were genetically modified to be herbicide-tolerant.

Formulations and Tradenames

Glyphosate is marketed in the United States and worldwide by many agrochemical companies, in different solution strengths and with various adjuvants, under dozens of tradenames. As of 2010, more than 750 glyphosate products were on the market. In 2012, in terms of volume about half of the total global consumption of glyphosate was for conventional crops; Asia Pacific was the largest and fastest growing market. Chinese manufacturers collectively are the world's largest producers of glyphosate and its precursors and account for about 30% of global exports. Key manufacturers include Anhui Huaxing Chemical Industry Company, BASF, Bayer CropScience, Dow AgroSciences, DuPont, Jiangsu Good Harvest-Weien Agrochemical Company, Monsanto, Nantong Jiangshan Agrochemical & Chemicals Co., Nufarm Limited, SinoHarvest, Syngenta, and Zhejiang Xinan Chemical Industrial Group Company.

Glyphosate is an acid molecule, so it is formulated as a salt for packaging and handling. Various salt formulations include isopropylamine, diammonium, monoammonium, or potassium as the counterion. Some brands include more than one salt. Some companies report their product as acid equivalent (ae) of glyphosate acid, or some report it as active ingredient (ai) of glyphosate plus the salt, and others report both. To compare performance of different formulations, knowledge of how the products were formulated is needed. Given that different salts have different weights, the acid equivalent is a more accurate method of expressing and comparing concentrations.

Adjuvant loading refers to the amount of adjuvant already added to the glyphosate product. Fully loaded products contain all the necessary adjuvants, including surfactant; some contain no adjuvant system, while other products contain only a limited amount of adjuvant (minimal or partial loading) and additional surfactants must be added to the spray tank before application. As of 2000 (just before Monsanto's patent on glyphosate expired), over 400 commercial adjuvants from over 34 different companies were available for use in commercial agriculture.

Products are supplied most commonly in formulations of 120, 240, 360, 480, and 680 g/l of active ingredient. The most common formulation in agriculture is 360 g/l, either alone or with added cationic surfactants.

For 360 g/l formulations, European regulations allow applications of up to 12 l/ha for control of perennial weeds such as couch grass. More commonly, rates of 3 l/ha are practiced for control of annual weeds between crops.

Monsanto

Monsanto's Roundup is the earliest formulation of glyphosate.

Monsanto developed and patented the use of glyphosate to kill weeds in the early 1970s, and first brought it to market in 1974, under the Roundup brandname. While its initial patent expired in 1991, Monsanto retained exclusive rights in the United States until its patent on the isopropylamine salt expired in September 2000.

As of 2009, sales of these herbicide products represented about 10% of Monsanto's revenue due to competition from other producers of other glyphosate-based herbicides; their Roundup products (which include GM seeds) represented about half of Monsanto's gross margin.

The active ingredient of the Monsanto herbicides is the isopropylamine salt of glyphosate. Another important ingredient in some formulations is the surfactant polyethoxylated tallow amine.

Monsanto also produces seeds which grow into plants genetically engineered to be tolerant to glyphosate. The genes contained in these seeds are patented. Such crops allow farmers to use glyphosate as a postemergence herbicide against most broadleaf and cereal weeds. Soy was the first glyphosate-resistant crop.

Toxicity

Glyphosate is the active ingredient in herbicide formulations containing it. However, in addition to glyphosate salts, commercial formulations of glyphosate contain additives such as surfactants which vary in nature and concentration. The surfactants are added to enable the glyphosate to penetrate the cuticle of the plants. Toxicologists have studied glyphosate alone and formulations.

Glyphosate Alone

Humans

Many regulatory and scholarly reviews have evaluated the relative toxicity of glyphosate as an herbicide. The German Federal Institute for Risk Assessment toxicology review in 2013 found that "the available data is contradictory and far from being convincing" with regard to correlations between exposure to glyphosate formulations and risk of various cancers, including non-Hodgkin lymphoma.

Early epidemiological studies did not find associations between long-term, low-level exposure to glyphosate and any disease. A 2000 review concluded that "under present and expected conditions of new use, there is no potential for Roundup herbicide to pose a health risk to humans". A 2002 review by the European Union reached the same conclusion. In 2013 the European commission reviewed a 2002 finding that had concluded equivocal evidence existed of a relationship between glyphosate exposure during pregnancy and cardiovascular malformations and found that "there is no increased risk at the levels of exposure below those that caused maternal toxicity." A 2013 review found that neither glyphosate nor typical glyphosate-based formulations pose a genotoxicity risk in humans under normal conditions of human or environmental exposures.

A 2014 review article reported a significant association between B-cell lymphoma and glyphosate occupational exposure. In March 2015, the World Health Organization's International Agency for Research on Cancer classified glyphosate as "probably carcinogenic in humans" (category 2A) based on epidemiological studies, animal studies, and *in vitro* studies. However, in 2016 a joint meeting of the United Nations (FAO) Panel of Experts on Pesticide Residues in Food and the Environment and the World Health Organization Core Assessment Group on Pesticide Residues concluded that based on the available evidence "glyphosate is unlikely to pose a carcinogenic risk to humans from exposure through the diet".

Other Mammals

Amongst mammals, glyphosate is considered to have "low to very low toxicity". The LD_{50} of glyphosate is 5,000 mg/kg for rats, 10,000 mg/kg in mice and 3,530 mg/kg in goats. The acute dermal LD_{50} in rabbits is greater than 2,000 mg/kg. Indications of glyphosate toxicity in animals typically appear within 30 to 120 minutes following ingestion of a large enough dose, and include initial excitability and tachycardia, ataxia, depression, and bradycardia, although severe toxicity can develop into collapse and convulsions.

A review of unpublished short-term rabbit-feeding studies reported severe toxicity effects at 150 mg/kg/day and "no observed adverse effect level" doses ranging from 50 to 200 mg/kg/day.

Glyphosate can have carcinogenic effects in nonhuman mammals. These include the induction of positive trends in the incidence of renal tubule carcinoma and haemangiosarcoma in male mice, and increased pancreatic islet-cell adenoma in male rats.

Glyphosate-based herbicides may cause life-threatening arrhythmias in mammals. Evidence also shows that such herbicides cause direct electrophysiological changes in the cardiovascular systems of rats and rabbits.

Other Aquatic Fauna

In freshwater invertebrates (species unspecified), glyphosate has a 48-hour LC_{50} ranging from 55 to 780 ppm. The 96-hour LC_{50} is 281 ppm for grass shrimp (*Palaemonetas vulgaris*) and 934 ppm for fiddler crabs (*Uca pagilator*). These values make glyphosate "slightly toxic to practically non-toxic".

Antimicrobial Activity

The antimicrobial activity of glyphosate has been described in the microbiology literature since its discovery in 1970 and the description of glyphosate's mechanism of action in 1972. Efficacy was described for numerous bacteria and fungi. Glyphosate can control the growth of apicomplexan parasites, such as *Toxoplasma gondii*, *Plasmodium falciparum* (malaria), and *Cryptosporidium parvum*, and has been considered an antimicrobial agent in mammals. Inhibition can occur with some Rhizobium species important for soybean nitrogen fixation, especially under moisture stress.

Soil Biota

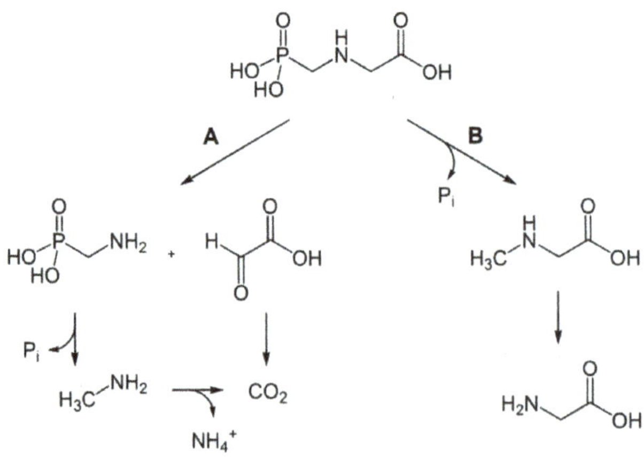

Degradation pathway of glyphosate in the ground

When glyphosate comes into contact with the soil, it can be bound to soil particles, thereby slowing its degradation. Glyphosate and its degradation product, aminomethylphosphonic acid, residues are considered to be much more benign toxicologically and environmentally than most of the herbicides replaced by glyphosate.

Other studies found that while "recommended dosages of glyphosate did not affect growth rates", much higher dosages reduced respiration in nitrogen-fixing bacteria.

Glyphosate can be harmful to some earthworms such as *Apporectodea caliginosa*, even at very low doses, and causes a significant reduction in growth of the earthworm *Eisenia fetida*.

Glyphosate-Based Formulations

Glyphosate-based formulations may contain a number of adjuvants, the identities of which are considered trade secrets. Surfactants are used in herbicide formulations as wetting agents, to maximize coverage and aid penetration of the herbicide(s) through plant leaves. As agricultural spray adjuvants, surfactants may be mixed into commercial formulations, such as Roundup, or they may be purchased separately and mixed on-site (tank mix).

Polyethoxylated tallow amine (POEA) is a surfactant used in the original Roundup formulation and was still being commonly used in 2015.Different versions of Roundup have included different percentages of POEA. Although Monsanto product fact sheets do not disclose surfactants and their percentages, a 1997 US government report said that Roundup is 15% POEA while Roundup Pro is 14.5%. A review of the literature provided to the EPA in 1997 found that POEA was more toxic to fish than glyphosate was. POEA is more toxic to fish and amphibians than glyphosate alone.

Human

A 2000 review concluded that "under present and expected conditions of new use, there is no potential for Roundup herbicide to pose a health risk to humans". A 2002 review by the European Union reached the same conclusion.

A 2012 meta-analysis of all epidemiological studies of exposure to glyphosate formulations found no correlation with any kind of cancer. The 2013 systematic review by the German Institute for Risk Assessment of epidemiological studies of workers who use pesticides, exposed to glyphosate formulations found no significant risk, stating that "the available data are contradictory and far from being convincing".[:Volume 1, p64-66]However, a 2014 meta-analysis of the same studies found a correlation between occupational exposure to glyphosate formulations and increased risk of B cell lymphoma, the most common kind of non-Hodgkin lymphoma. Workers exposed to glyphosate were about twice as likely to get B cell lymphoma.

Acute toxicity is dose-related; Skin exposure to ready-to-use glyphosate formulations can cause irritation, and photocontact dermatitis has been occasionally reported. These effects are probably due to the preservative benzisothiazolin-3-one. Severe skin burns are very rare. Inhalation is a minor route of exposure, but spray mist may cause oral or nasal discomfort, an unpleasant taste in the mouth, or tingling and irritation in the throat. Eye exposure may lead to mild conjunctivitis. Superficial corneal injury is possible if irrigation is delayed or inadequate. Death has been reported after deliberate overdose. Ingestion of Roundup ranging from 85 to 200 ml (of 41% solution) has resulted in death within hours of ingestion, although it has also been ingested in quantities as large as 500 ml with only mild or moderate symptoms. Consumption of over 85 ml of concentrated product are likely to cause serious symptoms in adults including burns due to corrosive effects as well as kidney and liver damage. More severe cases cause "respiratory distress, impaired consciousness, pulmonary edema, infiltration on chest X-ray, shock, arrhythmias, renal failure requiring haemodialysis, metabolic acidosis, and hyperkalaemia" and death is often preceeded by bradycardia and ventricular arrhythmias.

Other Animals

A 2000 review of the ecotoxicological data on Roundup shows at least 58 studies exist on the effects of Roundup on a range of organisms. This review concluded, "...for terrestrial uses of Roundup minimal acute and chronic risk was predicted for potentially exposed non-target organisms".

In reproductive toxicity studies performed in rats and rabbits, no adverse maternal or offspring effects were seen at doses below 175–293 mg/kg of body weight per day.

Fish

Monsanto and other companies produce glyphosate products with alternative surfactants specifically formulated for aquatic use, for example the Monsanto products "Biactive" and "AquaMaster". Glyphosate formulations are much more toxic for amphibians and fish than glyphosate alone. The half-life of POEA (21–42 days) is longer than that for glyphosate (7–14 days) in aquatic environments.

Amphibians

Some researchers have suggested the toxicity effects of pesticides on amphibians may be different from those of other aquatic fauna because of their lifestyle; amphibians may be more susceptible to the toxic effects of pesticides because they often prefer to breed in shallow, lentic, or ephemeral pools. These habitats do not necessarily constitute formal water-bodies and can contain higher concentrations of pesticide compared to larger water-bodies. Studies in a variety of amphibians have shown the toxicity of GBFs containing POEA to amphibian larvae. These effects include interference with gill morphology and mortality from either the loss of osmotic stability or asphyxiation. At sub-lethal concentrations, exposure to POEA or glyphosate/POEA formulations have been reported to be associated with delayed development, accelerated development, reduced size at metamorphosis, developmental malformations of the tail, mouth, eye and head, histological indications of intersex and symptoms of oxidative stress.

A 2003 study of various formulations of glyphosate found, "the risk assessments based on estimated and measured concentrations of glyphosate that would result from its use for the control of undesirable plants in wetlands and over-water situations showed that the risk to aquatic organisms is negligible or small at application rates less than 4 kg/ha and only slightly greater at application rates of 8 kg/ha."

A 2013 meta-analysis reviewed the available data related to potential impacts of glyphosate-based herbicides on amphibians. According to the authors, the use of glyphosate-based pesticides cannot be considered the major cause of amphibian decline, the bulk of which occurred prior to the widespread use of glyphosate or in pristine tropical areas with minimal glyphosate exposure. The authors recommended further study of species- and development-stage chronic toxicity, of environmental glyphosate levels, and ongoing analysis of data relevant to determining what if any role glyphosate might be playing in worldwide amphibian decline, and suggest including amphibians in standardized test batteries.

Other Aquatic Fauna

Glyphosate-based formulations can cause oxidative stress in bullfrog tadpoles and Pacific oysters.

Effect on Plant Health

A correlation was found between an increase in the infection rate of wheat by *Fusarium* head blight and the application of glyphosate, but "because of the nature of this study, we could not determine if the association between previous GF (glyphosate formulation) use and FHB development was a cause-effect relationship". Other studies have found causal relationships between glyphosate and decreased disease resistance. Exposure to glyphosate has been shown to change the species composition of endophytic bacteria in plant hosts, which is highly variable.

Endocrine Disruption

In 2007, the EPA selected glyphosate for further screening through its Endocrine Disruptor Screening Program (EDSP). Selection for this program is based on a compound's prevalence of use and does not imply particular suspicion of endocrine activity. On June 29, 2015 the EPA released Weight of Evidence Conclusion of the EDSP Tier 1 screening for glyphosate, recommending that glyphosate not be considered for Tier 2 testing. The Weight of Evidence conclusion stated "...there was no convincing evidence of potential interaction with the estrogen, androgen or thyroid pathways."

Genetic Damage

Several studies have not found mutagenic effects, so glyphosate has not been listed in the United States Environmental Protection Agency or the International Agency for Research on Cancer databases. Various other studies suggest glyphosate may be mutagenic. The IARC monograph noted that glyphosate-based formulations can cause DNA strand breaks in various taxa of animals *in vitro*

Government and Organization Positions

European Food Safety Authority

A 2013 systematic review by the German Institute for Risk Assessment (BfR) examined more than 1000 epidemiological studies, animal studies, and *in vitro* studies. It found that "no classification and labelling for carcinogenicity is warranted" and did not recommend a carcinogen classification of either 1A or 1B.[:139, 34–37] It provided the review to EFSA in January 2014 which published it in December 2014. On November, 12th, 2015, EFSA published its conclusion on the risk assessment of glyphosate, stating it was "unlikely to pose a carcinogenic hazard to humans".

EFSA's decision and the BfR report were criticized in an open letter published by 96 scientists in November 2015 saying that the BfR report failed to adhere to accepted scientific principles of open and transparent procedures. The BfR report included unpublished data, lacked authorship, omitted references, and did not disclose conflict-of-interest information.

On April 4, 2016, Dr. Vytenis Andriukaitis, European Commissioner for Health and Food Safety, wrote an open letter to the Chair of the Board of the Glyphosate Task at Monsanto Europe asking to publish the full studies provided to the EFSA.

US Environmental Protection Agency

The EPA, which last reviewed glyphosate in 1993, considers glyphosate to be noncarcinogenic and relatively low in dermal and oral acute toxicity. The EPA considered a "worst case" dietary risk model of an individual eating a lifetime of food derived entirely from glyphosate-sprayed fields with residues at their maximum levels. This model indicated that no adverse health effects would be expected under such conditions. In 2015, the EPA initiated a review glyphosate's toxicity and in 2016 reported their conclusion that glyphosate is likely not carcinogenic.

International Agency for Research on Cancer

In March 2015, the International Agency for Research on Cancer published a summary of their forthcoming monograph on glyphosate, and classified glyphosate as "probably carcinogenic in humans" (category 2A) based on epidemiological studies, animal studies, and *in vitro* studies. It noted that there was "limited evidence" of carcinogenicity in humans for non-Hodgkin lymphoma. The IARC classifies substances for their carcinogenic potential, and "a few positive findings can be enough to declare a hazard, even if there are negative studies, as well." Unlike the BfR, it does not conduct a so-called risk assessment weighing benefits against risk.

The BfR responded that IARC reviewed only a selection of what they had reviewed earlier, and argued that other studies, including a cohort study called ‹Agricultural Health Study', do not support the classification. The IARC report did not include the German regulatory study published in December 2014, nor did it include industry-funded studies Monsanto called the IARC report biased and said it wanted it to be retracted. It started a case against California's carcinogen classification in 2016.

Effects of Use

Emergence of Resistant Weeds

In the 1990s, when the first genetically modified crops-such as glyphosate-resistant corn, canola, soybean and cotton—were introduced,no glyphosate-resistant weeds existed. By 2014, glyphosate-resistant weeds dominated herbicide-resistant research. At that time, 23 glyphosate-resistant species were found in 18 countries.

"Resistance evolves after a weed population has been subjected to intense selection pressure in the form of repeated use of a single herbicide." Weeds resistant to the herbicide have been called '**superweeds**'.

In 2014, according to Ian Heap, a weed specialist who completed his PhD on resistance to multiple herbicides in annual ryegrass Lolium rigidum in 1988 (the first case of an herbicide-resistant weed in Australia), L. rigidum is the world's worst herbicide-resistant weed that affects over two million hectares in 12 countries, 11 sites of action, and 9 cropping regimens. Annual ryegrass was known to be resistant to herbicides since 1982. By 1996, the first documented case of glyphosate-resistant *L. rigidum* was reported in Australia in 1996 near Orange, New South Wales. In 2006, farmers associations were reporting 107 biotypes of weeds within 63 weed species with herbicide resistance. In 2009, Canada identified its first resistant weed, giant ragweed, and at that time 15 weed species had been confirmed as resistant to glyphosate. As of 2010, in the United States 7 to 10 million acres (2.8 to 4.0 million hectares) of soil were afflicted by superweeds, or about 5% of the 170 million acres planted with corn, soybeans, and cotton, the crops most affected, in 22 states. In 2012, Charles Benbrook reported that the Weed Science Society of America listed 22 superweeds in the U.S., with over 5.7×10^6 ha (14×10^6 acres) infested by GR weeds and that Dow AgroSciences had carried out a survey and reported a figure of around 40×10^6 ha (100×10^6 acres). The International Survey of Herbicide Resistant Weeds database lists species that are resistant to glyphosate.

In response to resistant weeds, farmers also perform hand-weeding, use tractors to turn over soil between crops, and use other herbicides in addition to glyphosate.

Monsanto scientists have found that some resistant weeds have as many as 160 extra copies of a gene called *EPSPS,* the enzyme glyphosate disrupts.

Palmer Amaranth

Amaranthus palmeri

In 2004, a glyphosate-resistant variation of *Amaranthus palmeri*, commonly known as Palmer amaranth, was found in Georgia and confirmed by a 2005 study. In 2005, resistance was also found in North Carolina. Widespread use of Roundup Ready crops led to an unprecedented selection pressure, and glyphosate resistance followed. The weed variation is now widespread in the southeastern United States. Cases have also been reported in Texas and Virginia.

Conyza

Conyza canadensis

Conyza bonariensis (also known as hairy fleabane and buva) and *Conyza canadensis* (known as horseweed or marestail), are other weed species that had lately developed glyphosate resistance. A 2008 study on the current situation of glyphosate resistance in South America concluded "resistance evolution followed intense glyphosate use" and the use of glyphosate-resistant soybean crops is a factor encouraging increases in glyphosate use. In the 2015 growing season, glyphosate-resistant marestail proved to be especially problematic to control in Nebraska production fields.

Ryegrass

Ryegrass Lolium perenne

Glyphosate-resistant ryegrass (*Lolium*) has occurred in most of the Australian agricultural areas and other areas of the world. All cases of evolution of resistance to glyphosate in Australia were characterized by intensive use of the herbicide while no other effective weed control practices were used. Studies indicate the resistant ryegrass does not compete well against nonresistant plants and their numbers decrease when not grown under conditions of glyphosate application.

Johnson Grass

Glyphosate-resistant Johnson grass (*Sorghum halepense*) is found in glyphosate-resistant soybean cultivation in northern Argentina.

Monarch Butterfly

Use of glyphosate to clear milkweed along roads and fields may have contributed to a decline in monarch butterfly populations in the Midwest. Along with deforestation and adverse weather conditions, the decrease in milkweed contributed to an 81% decline in monarchs. The Natural Resources Defense Council (NRDC) filed a suit in 2015 against the EPA, in which it is argued that the agency ignored warnings about the dangers of glyphosate usage for monarchs.

Legal Status

Glyphosate was first approved for use in the 1970s, and as of 2010 was labelled for use in 130 countries.[2]

In September 2013, the legislative assembly of El Salvador approved legislation to ban 53 agrochemicals, including glyphosate; the ban on glyphosate was set to begin in 2015.

In April 2014, the legislature of the Netherlands passed legislation prohibiting sale of glyphosate to individuals for use at home; commercial sales were not affected.

In May 2015, the president of Sri Lanka banned the use and import of glyphosate, effective immediately.

In May 2015, Bermuda blocked importation on all new orders of glyphosate-based herbicides for a temporary suspension awaiting outcomes of research.

In May 2015, Colombia announced that it would stop using glyphosate by October 2015 in the destruction of illegal plantations of coca, the raw ingredient for cocaine. Farmers have complained that the aerial fumigation has destroyed entire fields of coffee and other legal produce.

In June 2015, the French Ecology Minister asked nurseries and garden centers to halt over-the-counter sales of glyphosate in the form of Monsanto's Roundup. This was a nonbinding request and all sales of glyphosate remain legal in France until 2022, when the substance will be banned for home gardening.

A vote on the relicencing of glyphosate in the EU stalled in March 2016. Member states France, Sweden, and the Netherlands objected to the renewal. A vote to reauthorize on a temporary basis failed in June 2016 but at the last-minute the license was extended for 18 months and will be re-evaluated at the end of 2017.

Legal Cases

Advertising Controversy

The New York Times reported that in 1996, "Dennis C. Vacco, the Attorney General of New York, ordered the company Monsanto to pull ads that said Roundup was "safer than table salt" and "practically nontoxic" to mammals, birds and fish. The company withdrew the spots, but also said that the phrase in question was permissible under E.P.A. guidelines."

In 2001, French environmental and consumer rights campaigners brought a case against Monsanto for misleading the public about the environmental impact of its herbicide Roundup, on the basis that glyphosate, Roundup's main component, is classed as "dangerous for the environment" and "toxic for aquatic organisms" by the European Union. Monsanto's advertising for Roundup had presented it as biodegradable and as leaving the soil clean after use. In 2007, Monsanto was convicted of false advertising and was fined 15,000 euros. Monsanto's French distributor Scotts France was also fined 15,000 euros. Both defendants were ordered to pay damages of 5,000 euros to the Brittany Water and Rivers Association and 3,000 euros to the Consommation Logement Cadre de vie, one of the two main general consumer associations in France. Monsanto appealed and the court upheld the verdict; Monsanto appealed again to the French Supreme Court, and in 2009 it also upheld the verdict.

Trade Dumping Allegations

United States companies have cited trade issues with glyphosate being dumped into the western world market areas by Chinese companies and a formal dispute was filed in 2010.

REFERENCES

1. Erkut, Cihan (9 August 2011). "Trehalose Renders the Dauer Larva of Caenorhabditis elegans Resistant to Extreme Desiccation". Current Biology. 21 (15): 1331–1336. doi:10.1016/j.cub.2011.06.064. PMID 21782434. Retrieved 5 November 2014.
2. ^ Jump up to:a b Mattimore V, Battista JR (1996). "Radioresistance of Deinococcus radiodurans: functions necessary to survive ionizing radiation are also necessary to survive prolonged desiccation". J. Bacteriol. 178 (3): 633–7. PMC 177705 . PMID 8550493.
3. Zahradka K, Slade D, Bailone A, Sommer S, Averbeck D, Petranovic M, Lindner AB, Radman M (2006). "Reassembly of shattered chromosomes in Deinococcus radiodurans". Nature. 443 (7111): 569–73. doi:10.1038/nature05160. PMID 17006450.
4. Pitcher RS, Green AJ, Brzostek A, Korycka-Machala M, Dziadek J, Doherty AJ (2007). "NHEJ protects mycobacteria in stationary phase against the harmful effects of desiccation". DNA Repair (Amst.). 6 (9): 1271–6. doi:10.1016/j.dnarep.2007.02.009. PMID 17360246.
5. Dose K, Gill M (1995). "DNA stability and survival of Bacillus subtilis spores in extreme dryness". Orig Life Evol Biosph. 25 (1-3): 277–93. PMID 7708386.
6. Chown, S.L.,; Nicolson, S.W. (2004). "Insect Physiological Ecology". New York: Oxford University Press.
7. Gray, E.M.; Bradley, T.J. (2005). "Physiology of desiccation resistance in Anopheles gambiae and Anopheles arabiensis". American Journal of Tropical Medicine and Hygiene. 73 (3): 553–559. PMID 16172480.

8. Graves, J. L.; Toolson, E. C.; Jeong, C.; Vu & L.N.; Rose, M.R. (1992). "Desiccation, flight, glycogen and postponed senescence in Drosophila melanogaster". Physiological Zoology. 65: 268–286.
9. Gefen, E.; Marlon, A.J.; Gibbs, A.G. (2006). "Selection for desiccation resistance in adult Drosophila melanogaster affects larval development and metabolite accumulation". Journal of Experimental Biology. 209 (Pt 17): 3293–3300. doi:10.1242/jeb.02397. PMID 16916965.
10. Folk, D.G.; Han, C.; Bradley, T.J. (2001). "Water acquisition and partitioning in Drosophila melanogaster: effects of selection for desiccation-resistance". Journal of Experimental Biology. 204 (Pt 19): 3323–3331. PMID 11606606.
11. Mellanby, K. (1934). "The evaporation of water from insects". Journal of Experimental Biology. 10: 317–333. doi:10.1111/j.1469-185x.1935.tb00487.x.
12. Machin J. 1968. The Permeability of the Epiphragm of Terrestrial Snails to Water Vapor. Biological Bulletin, Vol. 134, No. 1 (Feb., 1968), pp. 87-95
13. Wigglesworth, V.B. (1984). "Insect Physiology". Chapman and Hall, NY.
14. Gibbs, A.G. (2002). "Lipid melting and cuticular permeability: new insights into an old problem". Journal of Insect Physiology. 48 (4): 391–400. doi:10.1016/S0022-1910(02)00059-8. PMID 12770088.
15. Treherne, J.E.; Willmer, P.G. (1975). "Hormonal control of integumentary water loss: evidence for a novel neuroendocrine system in an insect (Periplaneta americana)". Journal of Experimental Biology. 63 (1): 143–159. PMID 1159358.
16. Snyder, G.K.; Sheafor, B.; Scholnick, D.; Farrelly, C. (1995). "Gas Exchange in the Insect Tracheal System". Journal of Theoretical Biology. 172 (3): 199–207. doi:10.1006/jtbi.1995.0016. PMID 7715192.
17. Bursell, E. (1960). "Loss of water by excretion and defaecation in the Tsetse fly". Experimental Biology. 37: 689–697.
18. Willmer, P.; Stone, G.; Johnston, I.A. (2000). "Environmental physiology of animals". New York: Blackwell Publishing.
19. Renault, D.; Coray, Y. (2004). "Water loss of male and female Alphitobius diaperinus (Coleoptera: Tenebrionidae) maintained under dry conditions". European Journal of Entomology. 101: 491–494. doi:10.14411/eje.2004.069.
20. Prince, G.J.; Parsons, P.A. (1977). "Adaptive Behaviour of Drosophila Adults in Relation to Temperature and Humidity". Australian Journal of Zoology. 25 (2): 285–290. doi:10.1071/ZO9770285.
21. Halffter, G.; Edmonds, W.D. (1982). "The Nesting Behavior of Dung Beetles (Scarabaeinae) an Ecological and Evolutive Approach". Mexico D.F.: Instituto de Ecologia.
22. Rowley, M.; Hanson, F. (2007). "Humidity detection and hygropreference behavior in larvae of the tobacco hornworm, Manduca sexta". Journal of Insect Science. 7: 39. doi:10.1673/031.007.3901. PMC 2999434 . PMID 20302460.
23. Watanbe, M.; Kikawada, T.; Okuda, T. (2003). "Increase of internal ion concentration triggers trehalose synthesis associated with cryptobiosis in larvae

of Polypedilum vanderplanki". Journal of Experimental Biology. 206 (Pt 13): 2281–2286. doi:10.1242/jeb.00418. PMID 12771176.

24. "Assessment of large-scale test – Pre-harvest siccation in rape". Feiffer-consult. Retrieved 2 June 2013.
25. "The agronomic benefits of glyphosate in Europe" (PDF). Monsanto Europe SA. February 2010. Retrieved 2 June 2013.
26. "Desiccation programmes". Potato Council. 2 August 2011. Retrieved 2 June 2013.
27. http://www.fwi.co.uk/arable/novel-desiccant-targeted-for-stale-seed-bed-clean-up.htm
28. Brändli D, Reinacher S (January 2012). "Herbicides found in human urine". Ithaka Journal. Retrieved 3 June 2013.
29. Industry Task Force on Glyphosate (21 November 2013). "Preharvest weed control and desiccation.". Retrieved 15 January 2014.
30. Province of Saskatchewan Dept of Agriculture. October 2010 Herbicide Options to Enhance Harvesting FAQ
31. Arnason, Robert (28 May 2015). "Richardson Milling says glyphosate dessication sic acceptable for its oats". The Western Producer. Retrieved 8 June 2015.
32. Robert Harris for Farmers Weekly. June 29, 2012 Challenging season for desiccating oilseed rape
33. Lulai EC. Skin-set, Wound Healing, and Related Defects. Ch 22 in Potato Biology and Biotechnology: Advances and Perspectives: Advances and Perspectives Eds. Vreugdenhil D et al, Elsevier, 2011 ISBN 9780080525051
34. Kozlowski TT. Extend and Significance of Shedding Plant Parts. Chapter 1 in Physiological ecology Shedding of Plants Parts, Ed. T.T. Kozlowski. Elsevier, 2012 ISBN 9780323145602
35. Brian Hall (12 August 2014). "Direct Harvest: Do you need a desiccant or a pre-harvest glyphosate?". Field Crop News.
36. J. H. Orson and D. H. K. Davies (June 2007). "Pre-harvest glyphosate for weed control and as a harvest aid in cereals" (PDF). Home-Grown Cereals Authority.
37. Pete Berry (November 2006). "FSA Pesticide Residue Minimisation Crop Action Plan - Cereals" (PDF). Food Standards Agency.
38. E. D. Baxter, N. Byrd and I. R. Slaiding (November 2009). "Food safety review of UK cereal grain for use in malting, milling and animal feed" (PDF). Home-Grown Cereals Authority. p. 52.
39. Industry Task Force on Glyphosate (8 January 2014). "Preharvest use of glyphosate:Recent Austrian decision.". Retrieved 5 January 2014.
40. Arnason, Robert (22 April 2015). "Oat buyer says no glyphosate pre-harvest". The Western Producer. Retrieved 8 June 2015.
41. Daniels, Calvin (30 April 2015). "Oats on forefront in industry news". Carlyle Observer. Retrieved 10 June 2015.

42. Glyphosate, Environmental Health Criteria monograph No. 159, Geneva: World Health Organization, 1994, ISBN 92-4-157159-4
43. Index no. 607-315-00-8 of Annex VI, Part 3, to Regulation (EC) No 1272/2008 of the European Parliament and of the Council of 16 December 2008 on classification, labelling and packaging of substances and mixtures, amending and repealing Directives 67/548/EEC and 1999/45/EC, and amending Regulation (EC) No 1907/2006. OJEU L353, 31.12.2008, pp 1–1355 at pp 570, 1100..
44. US patent 3799758, Franz JE, "N-phosphonomethyl-glycine phytotoxicant compositions", issued 1974-03-26, assigned to Monsanto Company
45. United States EPA 2007 Pesticide Market Estimates Agriculture, Home and Garden
46. Myers JP, Antoniou MN, Blumberg B, Carroll L, Colborn T, Everett LG, Hansen M, Landrigan PJ, Lanphear BP, Mesnage R, Vandenberg LN, vom Saal FS, Welshons WV, Benbroo CM (February 17, 2016). "Concerns over use of glyphosate-based herbicides and risks associated with exposures: a consensus statement". 15 (19). Environmental Health: 13. doi:10.1186/s12940-016-0117-0. Retrieved March 24, 2016.
47. Sprankle P, Meggitt WF, Penner D (1975). "Rapid inactivation of glyphosate in the soil". Weed Science: 224–228.
48. "Glyphosate technical fact sheet (revised June 2015)". National Pesticide Information Center. 2010. Retrieved September 1, 2015.
49. "The agronomic benefits of glyphosate in Europe" (PDF). Monsanto Europe SA. February 2010. Retrieved 2013-06-02.
50. Cressey D (March 25, 2015). "Widely used herbicide linked to cancer". Nature. doi:10.1038/nature.2015.17181.
51. Renewal Assessment Report: Glyphosate. Volume 1. Report and Proposed Decision. December 18, 2013. German Institute for Risk Assessment, page 65. Downloaded from http://dar.efsa.europa.eu/dar-web/provision (registration required)
52. Schinasi L, Leon ME (Apr 2014). "Non-Hodgkin lymphoma and occupational exposure to agricultural pesticide chemical groups and active ingredients: a systematic review and meta-analysis". International Journal of Environmental Research and Public Health. 11 (4): 4449–527. doi:10.3390/ijerph110404449. PMC 4025008 . PMID 24762670.
53. ^ Jump up to:a b c d Guyton KZ, Loomis D, Grosse Y, El Ghissassi F, Benbrahim-Tallaa L, Guha N, Scoccianti C, Mattock H, Straif K (May 2015). "Carcinogenicity of tetrachlorvinphos, parathion, malathion, diazinon, and glyphosate". The Lancet. Oncology. 16 (5): 490–1. doi:10.1016/S1470-2045(15)70134-8. PMID 25801782.
54. "Press release: IARC Monographs Volume 112: evaluation of five organophosphate insecticides and herbicides" (PDF). International Agency for Research on Cancer, World Health Organization. March 20, 2015.
55. "European Food Safety Authority - Glyphosate report" (PDF). EFSA. Retrieved May 23, 2016.

56. “Glyphosate: EFSA updates toxicological profile | European Food Safety Authority”. www.efsa.europa.eu. Retrieved 2016-05-23.

57. “JOINT FAO/WHO MEETING ON PESTICIDE RESIDUES” (PDF). May 2016. Retrieved October 3, 2016.

58. Dill GM, Sammons RD, Feng PC, Kohn F, Kretzmer K, Mehrsheikh A, Bleeke M, Honegger JL, Farmer D, Wright D, Haupfear EA (2010). “Glyphosate: Discovery, Development, Applications, and Properties” (PDF). In Nandula VK. Glyphosate Resistance in Crops and Weeds: History, Development, and Management. Hoboken, N.J.: Wiley. ISBN 978-0-470-41031-8.

59. United States Patent 3,160,632 (1964) Stauffer Chemical

60. Alibhai MF, Stallings WC (Mar 2001). “Closing down on glyphosate inhibition — with a new structure for drug discovery”. Proceedings of the National Academy of Sciences of the United States of America. 98 (6): 2944–6. Bibcode:2001PNAS...98.2944A. doi:10.1073/pnas.061025898. JSTOR 3055165. PMC 33334 . PMID 11248008.

61. Monsanto’s John E. Franz Wins 1990 Perkin Medal Chem. Eng. News, 1990, 68 (11), pp. 29–30 doi:10.1021/cen-v068n011.p029

62. Pesticide Action Network UK.Glyphosate fact sheet Pesticides News No.33, September 1996, pp. 28-29 PAN-UK says it is “a welcome move away from chemicals which are highly toxic to humans and other nontarget organisms, and from chemicals which cause direct and lasting damage to the environment” and of course cautions against overuse.

63. “The National Medal of Technology and Innovation Recipients - 1987”. The United States Patent and Trademark Office. Retrieved 2012-11-29.

64. Stong C (May 1990). “People: Monsanto Scientist John E. Franz Wins 1990 Perkin Medal For Applied Chemistry”. The Scientist. 4 (10): 28.

65. Duke SO, Powles SB (2008), “Glyphosate: a once-in-a-century herbicide: Mini-review” (PDF), Pest Management Science, 64, pp. 319–325

66. Powles, Stephen B. (January 19, 2010). “Gene amplification delivers glyphosate-resistant weed evolution”. Proceedings of the National Academy of Sciences. pp. 955–956. doi:10.1073/pnas.0913433107. Retrieved March 24, 2016.

67. Chenier, Philip J. (2012). Survey of Industrial Chemistry (3rd ed.). Springer Science+Business Media. p. 384. ISBN 9781461506034.

68. Schuette J. “Environmental Fate of Glyphosate” (PDF). Department of Pesticide Regulation, State of California.

69. Zhou J, Li J, An R, Yuan H, Yu F (2012). “Study on a New Synthesis Approach of Glyphosate”. J. Agric. Food Chem. 60 (25): 6279–6285. doi:10.1021/jf301025p.

70. Steinrücken HC, Amrhein N (Jun 1980). “The herbicide glyphosate is a potent inhibitor of 5-enolpyruvyl-shikimic acid-3-phosphate synthase”. Biochemical and Biophysical Research Communications. 94 (4): 1207–12. doi:10.1016/0006-291X(80)90547-1. PMID 7396959.

71. Hock B, Elstner EF (28 September 2004). Plant Toxicology, Fourth Edition. CRC Press. pp. 292–296. ISBN 978-0-203-02388-4.
72. Purdue University, Department of Horticulture and Landscape Architecture, Metabolic Plant Physiology Lecture notes, Aromatic amino acid biosynthesis, The shikimate pathway – synthesis of chorismate.
73. Schönbrunn E, Eschenburg S, Shuttleworth WA, Schloss JV, Amrhein N, Evans JN, Kabsch W (Feb 2001). "Interaction of the herbicide glyphosate with its target enzyme 5-enolpyruvylshikimate 3-phosphate synthase in atomic detail". Proceedings of the National Academy of Sciences of the United States of America. 98 (4): 1376–80. Bibcode:2001PNAS...98.1376S. doi:10.1073/pnas.98.4.1376. PMC 29264 . PMID 11171958.
74. Glyphosate bound to proteins in the Protein Data Bank
75. Schulz A, Krüper A, Amrhein N (1985). "Differential sensitivity of bacterial 5-enolpyruvylshikimate-3-phosphate synthases to the herbicide glyphosate". FEMS Microbiology Letters. 28 (3): 297–301. doi:10.1111/j.1574-6968.1985.tb00809.x.
76. Pollegioni L, Schonbrunn E, Siehl D (Aug 2011). "Molecular basis of glyphosate resistance-different approaches through protein engineering". The FEBS Journal. 278 (16): 2753–66. doi:10.1111/j.1742-4658.2011.08214.x. PMC 3145815 . PMID 21668647.
77. Funke T, Han H, Healy-Fried ML, Fischer M, Schönbrunn E (Aug 2006). "Molecular basis for the herbicide resistance of Roundup Ready crops". Proceedings of the National Academy of Sciences of the United States of America. 103 (35): 13010–5. Bibcode:2006PNAS..10313010F. doi:10.1073/pnas.0603638103. JSTOR 30050705. PMC 1559744 . PMID 16916934.
78. Maeda H, Dudareva N (2012). "The shikimate pathway and aromatic amino Acid biosynthesis in plants". Annual Review of Plant Biology. 63: 73–105. doi:10.1146/annurev-arplant-042811-105439. PMID 22554242. The AAA pathways consist of the shikimate pathway (the prechorismate pathway) and individual postchorismate pathways leading to Trp, Phe, and Tyr.... These pathways are found in bacteria, fungi, plants, and some protists, but are absent in animals. Therefore, AAAs and some of their derivatives (vitamins) are essential nutrients in the human diet, although in animals Tyr can be synthesized from Phe by Phe hydroxylase....The absence of the AAA pathways in animals also makes these pathways attractive targets for antimicrobial agents and herbicides.
79. Borggaard, OK and AL Gimsing. 2008. Fate of glyphosate in soil and the possibility of leaching to ground and surface waters: a review. Pest Management Sci. 64:441–456.
80. Botta, F , G Lavisonb, G Couturier, F Alliot, E Moreau-Guigon, N Fauchon, B Guery, M Chevreuil, and H Blanchoud. 2009. Transfer of glyphosate and its degradate AMPA to surface waters through urban sewerage systems. Chemosphere 77 (1): 133–139.

81. Muniraa, S , A Farenhorsta, D Flatena, and C Grant. 2016. Phosphate fertilizer impacts on glyphosate sorption by soil. Chemosphere 153: 471–477.

82. Kanissery, RG, A Welsh, and GK Sims. 2014. Effect of soil aeration and phosphate addition on the microbial bioavailability of 14C-glyphosate. J. Environ. Quality 44:137-144. doi:10.2134/jeq2014.08.0331

83. "Registration Decision Fact Sheet for Glyphosate (EPA-738-F-93-011)" (PDF). R.E.D. FACTS. United States Environmental Protection Agency. 1993.

84. Borggaard OK, Gimsing AL (Apr 2008). "Fate of glyphosate in soil and the possibility of leaching to ground and surface waters: a review". Pest Management Science. 64 (4): 441–56. doi:10.1002/ps.1512. PMID 18161065.

85. ^ Jump up to:a b Andréa MM, Peres TB, Luchini LC, Bazarin S, Papini S, Matallo MB, Savoy VL (2003). "Influence of repeated applications of glyphosate on its persistence and soil bioactivity". Pesquisa Agropecuária Brasileira. 38 (11): 1329–1335. doi:10.1590/S0100-204X2003001100012.

86. Torstensson NT, Lundgren LN, Stenström J (Oct 1989). "Influence of climatic and edaphic factors on persistence of glyphosate and 2,4-D in forest soils". Ecotoxicology and Environmental Safety. 18 (2): 230–9. doi:10.1016/0147-6513(89)90084-5. PMID 2806176.

87. Albers CN, Banta GT, Hansen PE, Jacobsen OS (Oct 2009). "The influence of organic matter on sorption and fate of glyphosate in soil--comparing different soils and humic substances". Environmental Pollution. 157 (10): 2865–70. doi:10.1016/j.envpol.2009.04.004. PMID 19447533.

88. Ole K, Borggaard OK (2011). "Does phosphate affect soil sorption and degradation of glyphosate? - A review". Trends in Soil Science and Plant Nutrition. 2 (1): 17–27.

89. Sparling DW, Matson C, Bickham J, Doelling-Brown P (2006). "Toxicity of glyphosate as Glypro® and LI700 to red-eared slider (Trachemys scripta elegans) embryos and early hatchlings". Environmental Toxicology and Chemistry. 25 (10): 2768–2774. doi:10.1897/05-152.1.

90. Carey Gillam (Nov. 2016). "Alarming Levels of Glyphosate Found in Popular American Foods". EcoWatch.com. Retrieved 14 November 2016.

91. Amrein TM (2012-12-21). "Analysis of pesticides in food" (PDF). ETH Zurich. p. 15. Retrieved 2013-06-02.

92. Knezevic SZ (February 2010). "Use of Herbicide-Tolerant Crops as Part of an Integrated Weed Management Program". University of Nebraska Extension Integrated Weed Management Specialist. Archived from the original on June 15, 2010.

93. Nyamai PA, Prather TS, Wallace JM (2011). "Evaluating Restoration Methods across a Range of Plant Communities Dominated by Invasive Annual Grasses to Native Perennial Grasses". Invasive Plant Science and Management. 4: 306–316. doi:10.1614/IPSM-D-09-00048.1.

94. Luijendijk CD, Beltman WH, Smidt RA, van der Pas LJ, Kempenaar C (May

2005). "Measures to reduce glyphosate runoff from hard surfaces" (PDF). Plant Research International B.V. Wageningen.

95. Botta F, Lavison G, Couturier G, Alliot F, Moreau-Guigon E, Fauchon N, Guery B, Chevreuil M, Blanchoud H (Sep 2009). "Transfer of glyphosate and its degradate AMPA to surface waters through urban sewerage systems". Chemosphere. 77 (1): 133–9. doi:10.1016/j.chemosphere.2009.05.008. PMID 19482331.
96. BBC. May 10, 2015. Colombia to ban coca spraying herbicide glyphosate
97. "Sugarcane Ripener Recommendations - Glyphosate". Louisiana State University Agricultural Extension Office. 3 September 2014. Retrieved 3 September 2014.
98. Green JM, Owen MD (Jun 2011). "Herbicide-resistant crops: utilities and limitations for herbicide-resistant weed management". Journal of Agricultural and Food Chemistry. 59 (11): 5819–29. doi:10.1021/jf101286h. PMC 3105486 . PMID 20586458.
99. Rashid A (2009). Introduction to Genetic Engineering of Crop Plants: Aims and Achievements. I K International. p. 259. ISBN 978-93-80026-16-9.
100. "Company History". Web Site. Monsanto Company.
101. "Adoption of Genetically Engineered Crops in the U.S.". Economic Research Service. USDA. Retrieved 26 August 2015.
102. Farm Chemicals International Glyphosate entry in Crop Protection Database
103. Alberta Agriculture and Rural Development. April 26, 2006. Quick Guide to Glyphosate Products - Frequently Asked Questions
104. Hartzler, Bob. "Glyphosate: a Review". ISU Weed Science Online. Iowa State University Extension.
105. Tu M, Hurd C, Robison R, Randall JM (November 1, 2001). "Glyphosate" (PDF). Weed Control Methods Handbook. The Nature Conservancy.
106. National Pesticide Information Center. Last updated September 2010 Glyphosate General Fact Sheet
107. Reuters. Apr 30, 2014. Press Release: Research and Markets: Global Glyphosate Market for Genetically Modified and Conventional Crops 2013 - 2019
108. China Research & Intelligence, June 5, 2013. Research Report on Global and China Glyphosate Industry, 2013-2017
109. Tu M, Randall JM (2003-06-01). "Glyphosate" (PDF). Weed Control Methods Handbook. The Nature Conservancy.
110. Curran WS, McGlamery MD, Liebl RA, Lingenfelter DD (1999). "Adjuvants for Enhancing Herbicide Performance". Penn State Extension.
111. VanGessel M. "Glyphosate Formulations". Control Methods Handbook, Chapter 8, Adjuvants: Weekly Crop Update. University of Delaware Cooperative Extension.
112. Sprague C, Hager A (2000-05-12). "Principles of Postemergence Herbicides". University of Illinois Extension Service. Retrieved 2012-11-29.
113. Young B. "Adjuvant Products by Manufacturer, Compendium of Herbicide Adjuvants". Southern Illinois University.

114. e-phy: Le catalogue des produits phytopharmaceutiques et de leurs usages des matières fertilisantes et des supports de culture homologués en France
115. "History of Monsanto's Glyphosate Herbicides" (PDF). Monsanto. Retrieved 20 December 2015.
116. US application 3799758, John E. Franz, "N-phosphonomethyl-glycine phytotoxicant compositions", published 1974-03-26, assigned to Monsanto Company
117. US application 4405531, John E. Franz, "Salts of N-phosphonomethylglycine", published 1983-09-20, assigned to Monsanto Company
118. Fernandez, Ivan (2002-05-15). "The Glyphosate Market: A 'Roundup'". Frost & Sullivan. Archived from the original on 2016-03-04. Retrieved 2015-03-10.
119. "The debate over whether Monsanto is a corporate sinner or saint". The Economist. 19 November 2009. Retrieved 20 November 2009.
120. Cavallaro M (2009-06-26). "The Seeds Of A Monsanto Short Play". Forbes. Retrieved 2009-07-11.
121. Mink PJ, Mandel JS, Lundin JI, Sceurman BK (Nov 2011). "Epidemiologic studies of glyphosate and non-cancer health outcomes: a review". Regulatory Toxicology and Pharmacology. 61 (2): 172–84. doi:10.1016/j.yrtph.2011.07.006. PMID 21798302.
122. Mink PJ, Mandel JS, Sceurman BK, Lundin JI (Aug 2012). "Epidemiologic studies of glyphosate and cancer: a review". Regulatory Toxicology and Pharmacology. 63 (3): 440–52. doi:10.1016/j.yrtph.2012.05.012. PMID 22683395.
123. Williams AL, Watson RE, DeSesso JM (2012). "Developmental and reproductive outcomes in humans and animals after glyphosate exposure: a critical analysis". Journal of Toxicology and Environmental Health. Part B, Critical Reviews. 15 (1): 39–96. doi:10.1080/10937404.2012.632361. PMID 22202229.
124. Williams GM, Kroes R, Munro IC (Apr 2000). "Safety evaluation and risk assessment of the herbicide Roundup and its active ingredient, glyphosate, for humans". Regulatory Toxicology and Pharmacology. 31 (2 Pt 1): 117–65. doi:10.1006/rtph.1999.1371. PMID 10854122.
125. "Review report for the active substance glyphosate" (PDF). Commission working document. European Commission, Health and Protection Directorate-General: Directorate E – Food Safety: plant health, animal health and welfare, international questions: E1 - Plant Health. 2002-01-21.
126. Kimmel GL, Kimmel CA, Williams AL, DeSesso JM (Feb 2013). "Evaluation of developmental toxicity studies of glyphosate with attention to cardiovascular development". Critical Reviews in Toxicology. 43 (2): 79–95. doi:10.3109/10408444.2012.749834. PMC 3581053 . PMID 23286529.
127. Kier LD, Kirkland DJ (Apr 2013). "Review of genotoxicity studies of glyphosate and glyphosate-based formulations". Critical Reviews in Toxicology. 43 (4): 283–315. doi:10.3109/10408444.2013.770820. PMID 23480780.

128. "JOINT FAO/WHO MEETING ON PESTICIDE RESIDUES" (PDF). May 2016. Retrieved June 2016. Check date values in: |access-date=(help)
129. Kimmel, Gary (2013). "Evaluation of developmental toxicity studies of glyphosate with attention to cardiovascular development". Critical Reviews in Toxicology. 43 (2): 79–95. doi:10.3109/10408444.2012.749834. PMC 3581053 . PMID 23286529.
130. Gress S, Lemoine S, Séralini GE, Puddu PE (April 2015). "Glyphosate-based herbicides potently affect cardiovascular system in mammals: review of the literature". Cardiovascular Toxicology. 15 (2): 117–26. doi:10.1007/s12012-014-9282-y. PMID 25245870.
131. Abraham William (Wildwood. Glyphosate formulations and their use for inhibition of 5-enolpyrovylshikimate-3-phosphate synthase, US Patent 7, 771, 736 B2; 2010
132. Roberts CW, Roberts F, Lyons RE, Kirisits MJ, Mui EJ, Finnerty J, Johnson JJ, Ferguson DJ, Coggins JR, Krell T, Coombs GH, Milhous WK, Kyle DE, Tzipori S, Barnwell J, Dame JB, Carlton J, McLeod R (Feb 2002). "The shikimate pathway and its branches in apicomplexan parasites". The Journal of Infectious Diseases. 185 (Suppl 1): S25–36. doi:10.1086/338004. PMID 11865437.
133. Zablotowicz RM, Reddy KN (2004). "Impact of glyphosate on the Bradyrhizobium japonicum symbiosis with glyphosate-resistant transgenic soybean: a minireview". Journal of Environmental Quality. 33 (3): 825–31. doi:10.2134/jeq2004.0825. PMID 15224916.
134. Giesy JP, Dobson S, Solomon KR (2000). "Ecotoxicological risk assessment for Roundup® herbicide". Reviews of Environmental Contamination and Toxicology. 167: 35–120. doi:10.1007/978-1-4612-1156-3_2. ISBN 978-0-387-95102-7.
135. Cerdeira AL, Duke SO (January 2010). "Effects of glyphosate-resistant crop cultivation on soil and water quality". GM Crops. 1 (1): 16–24. doi:10.4161/gmcr.1.1.9404. PMID 21912208.
136. Santos A, Flores M (1995). "Effects of glyphosate on nitrogen fixation of free-living heterotrophic bacteria". Letters in Applied Microbiology. 20 (6): 349–52. doi:10.1111/j.1472-765X.1995.tb01318.x.
137. Yamada T, Kremer RJ, de Camargo e Castro PR, Wood BW (2009). "Glyphosate interactions with physiology, nutrition, and diseases of plants: Threat to agricultural sustainability?". European Journal of Agronomy. Elsevier. 31 (3): 111–113. doi:10.1016/j.eja.2009.07.004.
138. Paoletti MG (1999). "The role of earthworms for assessment of sustainability and as bioindicators". In Paoletti MG. Invertebrate Biodiversity as Bioindicators of Sustainable Landscapes: Practical Use of Invertebrates to Assess Sustainable Land Use. Elsevier. pp. 137–156.
139. Yasmin S, D'Souza D (2010). "Effects of pesticides on the growth and reproduction of earthworm: a review". Applied and Environmental Soil Science. 2010: 1–9. doi:10.1155/2010/678360.
140. "Pesticide Registration Manual | Pesticide Registration | US EPA".

141. "Measuring POEA, a Surfactant Mixture in Herbicide Formulations". U.S. Geological Survey.
142. Gary L. Diamond and Patrick R. Durkin February 6, 1997, under contract from the United States Department of Agriculture. Effects of Surfactants on the Toxicity of Glyphosate, with Specific Reference to RODEO
143. Mann RM, Hyne RV, Choung CB, Wilson SP (2009). "Amphibians and agricultural chemicals: Review of the risks in a complex environment". Environmental Pollution. 157 (11): 2903–2927. doi:10.1016/j.envpol.2009.05.015.
144. Bradberry SM, Proudfoot AT, Vale JA (2004). "Glyphosate poisoning". Toxicological Reviews. 23 (3): 159–67. doi:10.2165/00139709-200423030-00003. PMID 15862083.
145. Sribanditmongkol P, Jutavijittum P, Pongraveevongsa P, Wunnapuk K, Durongkadech P (Sep 2012). "Pathological and toxicological findings in glyphosate-surfactant herbicide fatality: a case report". The American Journal of Forensic Medicine and Pathology. 33 (3): 234–7. doi:10.1097/PAF.0b013e31824b936c. PMID 22835958.
146. Talbot AR, Shiaw MH, Huang JS, Yang SF, Goo TS, Wang SH, Chen CL, Sanford TR (Jan 1991). "Acute poisoning with a glyphosate-surfactant herbicide ('Roundup'): a review of 93 cases". Human & Experimental Toxicology. 10 (1): 1–8. doi:10.1177/096032719101000101. PMID 1673618.
147. "Response to "The impact of insecticides and herbicides on the biodiversity and productivity of aquatic communities"" (PDF). Backgrounder. Monsanto Company. 2005-04-01.
148. "Aquatic Use of Glyphosate Herbicides in Australia" (PDF). Backgrounder. Monsanto Company. 2003-05-01.
149. Mesnage R, Defarge N, Spiroux de Vendômois J, Séralini GE (2015). "Potential toxic effects of glyphosate and its commercial formulations below regulatory limits". Food Chem. Toxicol. 84: 133–53. doi:10.1016/j.fct.2015.08.012. PMID 26282372.
150. Govindarajulu PP (2008). "Literature Review of Impacts of Glyphosate Herbicide on Amphibians: What Risks can the Silvicultural Use of this Herbicide Pose for Amphibians in BC?". British Columbia, Ecosystems Branch, Ministry of Environment. Retrieved December 12, 2015.
151. Solomon KR, Thompson DG (2003). "Ecological risk assessment for aquatic organisms from over-water uses of glyphosate". Journal of Toxicology and Environmental Health. Part B, Critical Reviews. 6 (3): 289–324. doi:10.1080/10937400306468. PMID 12746143.
152. Wagner N, Reichenbecher W, Teichmann H, Tappeser B, Lötters S (Aug 2013). "Questions concerning the potential impact of glyphosate-based herbicides on amphibians". Environmental Toxicology and Chemistry / SETAC. 32 (8): 1688–700. doi:10.1002/etc.2268. PMID 23637092.
153. "IARC monograph on glyphosate" (PDF). IARC. Archived from the original (PDF) on 2015-09-05. Retrieved September 27, 2015.

154. Fernandez MR, Selles F, Gehl D, Depauw RM, Zentner RP (2005). "Crop Production Factors Associated with Fusarium Head Blight in Spring Wheat in Eastern Saskatchewan". Crop Science. 45 (5): 1908–16. doi:10.2135/cropsci2004.0197.

155. Duke SO, Wedge DE, Cerdeira AL, Matallo MB (2007). "Interactions of Synthetic Herbicides with Plant Disease and Microbial Herbicides". In Vurro M, Gressel J. Novel Biotechnologies for Biocontrol Agent Enhancement and Management. NATO Security through Science Series. pp. 277–96. doi:10.1007/978-1-4020-5799-1_15. ISBN 978-1-4020-5797-7.

156. Rosenblueth M, Martínez-Romero E (Aug 2006). "Bacterial endophytes and their interactions with hosts". Molecular Plant-Microbe Interactions. 19 (8): 827–37. doi:10.1094/MPMI-19-0827. PMID 16903349.

157. United States Environmental Protection Agency (18 June 2007). "Draft List of Initial Pesticide Active Ingredients and Pesticide Inerts to be Considered for Screening under the Federal Food, Drug, and Cosmetic Act" (PDF). Federal Register. 72 (116): 33486–503.

158. United States Environmental Protection Agency (29 June 2015). "Memorandum: EDSP Weight of Evidence Conclusions on the Tier 1 Screening Assays for the List 1 Chemicals" (PDF).

159. ToxNet. Glyposate. National Library of Medicine.

160. András Székács and Béla Darvas. Forty years with glyphosate. In: Herbicides - Properties, Synthesis and Control of Weeds", Ed. Mohammed Naguib Abd El-Ghany Hasaneen, ISBN 978-953-307-803-8, Published: January 13, 2012.

161. http://www.bfr.bund.de/en/press_information/2014/03/glyphosate__no_more_poisonous_than_previously_assumed__although_a_critical_view_should_be_taken_of_certain_co_formulants-188898.html

162. "Glyphosate RAR 01 Volume 1 2013-12-18 San". Renewal Assessment Report. Hungry4Pesticides. 18 December 2013. Retrieved 27 March 2015.

163. Bundesinstitut für Risikobewertung. Updated 15 January 2014 Frequently asked questions on the health assessment of glyphosate

164. "Conclusion on the peer review of the pesticide risk assessment of the active substance glyphosate" (PDF). EFSA Journal. 13 (11): 4302. 2015. doi:10.2903/j.efsa.2015.4302.

165. "Independent scientists warn over Monsanto herbicide". DW. December 1, 2015. Retrieved December 9, 2015.

166. Portier CJ, et al. (27 November 2015). "Open letter: Review of the Carcinogenicity of Glyphosate by EFSA and BfR" (PDF). Retrieved 9 December 2015.

167. "Open letter. Subject: Plant protection products - transparency in the context of the decision-making process on glyphosate" (PDF). European Commission. 4 April 2016. Retrieved 16 May 2016.

168. Charles, Dan (17 September 2016). "EPA Weighs In On Glyphosate, Says It Doesn't Cause Cancer". NPR. Retrieved 19 September 2016.

169. Specter M (10 April 2015). "Roundup and Risk Assessment". New Yorker. 'Probable' means that there was enough evidence to say it is more than possible, but not enough evidence to say it is a carcinogen," Aaron Blair, a lead researcher on the IARC's study, said. Blair, a scientist emeritus at the National Cancer Institute, has studied the effects of pesticides for years. "It means you ought to be a little concerned about" glyphosate, he said.

170. Pollack A (27 March 2015). "Weed Killer, Long Cleared, Is Doubted". New York Times.

171. "Löst glyphosat Krebs aus? (announcement 007/2015)" (PDF) (in German). German Institute for Risk Assessment. 23 March 2015.

172. Gillam C (24 March 2015). "Monsanto seeks retraction for report linking herbicide to cancer". Reuters.

173. Reuters (2016-01-21). "Monsanto Sues California Over Herbicide Classification". The New York Times. ISSN 0362-4331. Retrieved 2016-01-25.

174. James, Clive (1996). "Global Review of the Field Testing and Commercialization of Transgenic Plants: 1986 to 1995" (PDF). The International Service for the Acquisition of Agri-biotech Applications. Retrieved July 17, 2010.

175. "Roundup Ready System". Monsanto.

176. "Resisting Roundup". The New York Times. May 16, 2010. Retrieved March 24, 2016.

177. Heap, Ian (2014-04-11). Herbicide Resistant Weeds. Integrated Pest Management. doi:10.1007/978-94-007-7796-5_12. Retrieved 2016-03-24.

178. Lori (2009-05-07). "U of G Researchers Find Suspected Glyphosate-Resistant Weed". Uoguelph.ca. Retrieved 2010-08-22.

179. Tarter S (2009-04-06). "Attack of the Superweeds". Journal Star (Peoria). Retrieved 2010-08-22.

180. Heap, Ian Michael (1988). "Resistance to herbicides in annual ryegrass (Lolium Rigidum)". Adelaide: Department of Agronomy, University of Adelaide.

181. King, Carolyn (June 2015). "History of herbicide resistance Herbicide resistance: Then, now, and the years to come". Retrieved 24 March 2016.

182. Hartzler, Bob (January 29, 2003), "Are RR Weeds in Your Future I", Iowa State University (ISU), Weed Science Online, retrieved March 24, 2016

183. Powles SB, Lorraine-Colwill DF, Dellow JJ, Preston C (1998). "Evolved Resistance to Glyphosate in Rigid Ryegrass (Lolium rigidum) in Australia". Weed Science. 46 (5): 604–7. JSTOR 4045968.

184. Glyphosate Resistance in Crops and Weeds: History, Development, and Management. Editor, Vijay K. Nandula. John Wiley & Sons, 2010 ISBN 9781118043547

185. "Glyphosate resistance is a reality that should scare some cotton growers into changing the way they do business". Southeastfarmpress.com. Retrieved 2010-08-22.

186. "Map of Glyphosate-Resistant Weeds Globally". The International Survey of Herbicide Resistant Weeds. 2010. Retrieved 12 Jan 2013.

187. Neuman W, Pollack A (4 May 2010). "U.S. Farmers Cope With Roundup-Resistant Weeds". New York Times. New York. pp. B1. Retrieved 4 May 2010.

188. Benbrook CM (2012). "Impacts of genetically engineered crops on pesticide use in the U.S. - the first sixteen years". Environmental Sciences Europe. 24: 24. doi:10.1186/2190-4715-24-24.

189. Heap I. "Resistance by Active Ingredient (select "glyphosate" from the pulldown menu)". The International Survey of Herbicide Resistant Weeds. Retrieved 13 April 2014.

190. "With BioDirect, Monsanto Hopes RNA Sprays Can Someday Deliver Drought Tolerance and Other Traits to Plants on Demand". MIT Technology Review. Retrieved 2015-08-31.

191. Culpepper AS, Grey TL, Vencill WK, Kichler JM, Webster TM, Brown SM, York AC, Davis JW, Hanna WW (2006). "Glyphosate-resistant Palmer amaranth (Amaranthus palmeri) confirmed in Georgia". Weed Science. 54 (4): 620–6. doi:10.1614/WS-06-001R.1. JSTOR 4539441.

192. Hampton N. "Cotton versus the monster weed". Retrieved 2009-07-19.

193. Smith JT (March 2009). "Resistance a growing problem" (PDF). The Farmer Stockman. Retrieved 2009-07-19.

194. Taylor O (2009-07-16). "Peanuts: variable insects, variable weather, Roundup resistant Palmer in new state". PeanutFax. AgFax Media. Retrieved 2009-07-19.

195. Vargas L, Bianchi MA, Rizzardi MA, Agostinetto D, Dal Magro T (2007). "Buva (Conyza bonariensis) resistente ao glyphosate na região sul do Brasil" Conyza bonariensis biotypes resistant to the glyphosate in southern Brazil. Planta Daninha (in Portuguese). 25 (3): 573–8. doi:10.1590/S0100-83582007000300017.

196. Koger CH, Shaner DL, Henry WB, Nadler-Hassar T, Thomas WE, Wilcut JW (2005). "Assessment of two nondestructive assays for detecting glyphosate resistance in horseweed (Conyza canadensis)". Weed Science. 53 (4): 438–45. doi:10.1614/WS-05-010R. JSTOR 4047050.

197. Ge X, d'Avignon DA, Ackerman JJ, Sammons RD (Apr 2010). "Rapid vacuolar sequestration: the horseweed glyphosate resistance mechanism". Pest Management Science. 66 (4): 345–8. doi:10.1002/ps.1911. PMC 3080097 . PMID 20063320.

198. Vila-Aiub MM, Vidal RA, Balbi MC, Gundel PE, Trucco F, Ghersa CM (Apr 2008). "Glyphosate-resistant weeds of South American cropping systems: an overview". Pest Management Science. 64 (4): 366–71. doi:10.1002/ps.1488. PMID 18161884.

199. Jhala A (4 June 2015). "Post-Emergence Herbicide Options for Glyphosate-Resistant Marestail in Corn and Soybean". CropWatch. Nebraska Extension. Retrieved 17 August 2015.

200. Preston C, Wakelin AM, Dolman FC, Bostamam Y, Boutsalis P (2009). "A Decade of Glyphosate-Resistant Lolium around the World: Mechanisms, Genes, Fitness,

and Agronomic Management". Weed Science. 57 (4): 435–41. doi:10.1614/WS-08-181.1.

201. Vila-Aiub MM, Balbi MC, Gundel PE, Ghersa CM, Powles SB (2007). "Evolution of Glyphosate-Resistant Johnsongrass (Sorghum halepense) in Glyphosate-Resistant Soybean". Weed Science. 55 (6): 566–71. doi:10.1614/WS-07-053.1. JSTOR 4539618.
202. Kniss, Andrew (2014-02-10). "Are herbicides responsible for the decline in Monarch butterflies?". Control Freaks. Retrieved 2016-06-16. The evidence seems clear that the number of milkweed plants through this region has indeed declined. The cause for the milkweed decline, though, is a little less certain.
203. Plumer, Brad (January 29, 2014). "Monarch butterflies keep disappearing. Here's why.". The Washington Post. Retrieved 2016-06-16.
204. Pleasants JM, Oberhauser KS (2012). "Milkweed loss in agricultural fields because of herbicide use: effect on the monarch butterfly population" (PDF). Insect Conservation and Diversity. 6: 135–144. doi:10.1111/j.1752-4598.2012.00196.x.
205. Hartzler RG, Buhler DD (2000). "Occurrence of common milkweed (Asclepias syriaca) in cropland and adjacent areas". Crop Protection. 19: 363–366. doi:10.1016/s0261-2194(00)00024-7.
206. "NRDC Sues EPA Over Demise of Monarch Butterfly Population". NBC. 2015.
207. Staff, Centralamericadata.com. September 6, 2013 El Salvador: Use of 53 Chemicals Banned
208. Staff, Centralamericadata.com. November 27, 2013 El Salvador: Confirmation to Be Given on Ban of Agrochemicals
209. Legislative Assembly of El Salvador. November 26, 2013 Analizan observaciones del Ejecutivo al decreto que contiene la prohibición de 53 agroquímicos que dañan la salud English translation by Google
210. Staff, Sustainable Pulse. Apr 4 2014 Dutch Parliament Bans Glyphosate Herbicides for Non-Commercial Use
211. Staff, Colombo Page. May 22, 2015 Sri Lankan President orders to ban import of glyphosate with immediate effect
212. Sarina Locke for the Australian Broadcasting Corporation. Updated May 27, 2015 Toxicologist critical of 'dodgy science' in glyphosate bans
213. "Health Minister: importation of roundup weed spray suspended". Bermuda Today. 11 May 2015.
214. "Colombia to ban coca spraying herbicide glyphosate". BBC News. 10 May 2015.
215. "UPDATE 2-French minister asks shops to stop selling Monsanto Roundup weedkiller". Reuters.
216. Arthur Nelson (2016-03-08). "Vote on controversial weedkiller's European licence postponed". The Guardian.
217. "Recall of Monsanto's Roundup likely as EU refuses limited use of glyphosate". Reuters. 2016-06-06.

218. Arthur Nelson (2016-06-29). "Controversial chemical in Roundup weedkiller escapes immediate ban". The Guardian.
219. Charry T (1997-05-29). "Monsanto recruits the horticulturist of the San Diego Zoo to pitch its popular herbicide". Business Day. New York Times.
220. "Monsanto fined in France for 'false' herbicide ads". Terradaily.com (January 26, 2007).
221. Europe | Monsanto guilty in 'false ad' row. BBC News (October 15, 2009).
222. Piller D (2010-04-01). "Albaugh accuses Chinese of dumping herbicide". Staff Blogs. Des Moines Register.
223. "In the Matter of: GLYPHOSATE FROM CHINA" (PDF). United States International Trade Commission. 2010-04-22.

CHAPTER 6

Salt Tolerance Plants

Chapter Contents

A halophyte is a plant which is capable of surviving in a highly salty environment. An estimated two percent of plant species are halophytic, with the vast majority of plants being glycophytes which will not survive on a salty substrate. Although halophytes make up a very small percentage of the overall plant population, they play a number of important roles in the environment, and they have some potentially useful industrial applications.

INTRODUCTION

A halophyte is a plant that grows in waters of high salinity, coming into contact with saline water through its roots or by salt spray, such as in saline semi-deserts, mangrove swamps, marshes and sloughs and seashores. An example of a halophyte is the salt marsh grass Spartina alterniflora (smooth cordgrass). Relatively few plant species are halophytes—perhaps only 2% of all plant species.

The large majority of plant species are glycophytes, which are not salt-tolerant and are damaged fairly easily by high salinity.

Spartina alterniflora (cordgrass), a halophyte.

Salt Tolerance

One quantitative measure of salt tolerance (halotolerance) is the total dissolved solids in irrigation water that a plant can tolerate. Seawater typically contains 40 grams per litre (g/l) of dissolved salts (mostly sodium chloride). Beans and rice can tolerate about 1–3 g/l, and are considered glycophytes (as are most crop plants). At the other extreme, *Salicornia bigelovii* (dwarf glasswort) grows well at 70 g/l of dissolved solids, and is a promising halophyte for use as a crop. Plants such as barley (*Hordeum vulgare*) and the date palm (*Phoenix dactylifera*) can tolerate about 5 g/l, and can be considered as marginal halophytes.

Adaptation to saline environments by halophytes may take the form of salt tolerance or salt avoidance. Plants that avoid the effects of high salt even though they live in a saline environment may be referred to as facultative halophytes rather than 'true', or obligatory, halophytes.

For example, a short-lived plant species that completes its reproductive life cycle during periods (such as a rainy season) when the salt concentration is low would be avoiding salt rather than tolerating it. Or a plant species may maintain a 'normal' internal salt concentration by excreting excess salts through its leaves, by way of a hydathode, or by concentrating salts in leaves that later die and drop off. In an effort to improve agricultural production in regions where crops are exposed to salinity, research is focused on improving understanding of the various mechanisms whereby plants respond to salinity stress, so that more robust crop halophytes may be developed. Adaptive responses to salinity stress have been identified at molecular, cellular, metabolic, and physiological levels.

ANEMOPSIS

The monotypic genus *Anemopsis* has only one species, *Anemopsis californica*, with the common names yerba mansa or lizard tail.

It is a perennial herb in the lizard tail family (Saururaceae) and prefers very wet soil or shallow water.

Range and Habitat

It is native to southwestern North America in Northwest Mexico and the Southwestern United States from California to Texas to Kansas to Oregon.

Description

Leaves and Stems

As it matures, the visible part of the plant develops red stains, eventually turning bright red in the fall.

Inflorescence and Fruit

Yerba mansa is showy in spring when in bloom. The iconic white "flowers" (actually reduced inflorescences, or pseudanthia) are borne in early spring, and are surrounded by 4–9 large white bracts.

Similar to the sunflower family, what appears to be a single bloom is in reality a dense cluster of individually small flowers borne in an inflorescence. In this species the inflorescence is conical and has five to ten large white bracts beneath it, so that along with the tiny white florets, the whole structure is quite striking when it blooms in spring. The conical structure develops into a single, tough fruit that can be carried downstream to spread the tiny, pepper-like seeds.

Common Name

In her book on herbs of the southwestern USA, Dr. Soule discusses the common name. "*Yerba mansa* is one of those names which confounds linguists. *Yerba* is Spanish for herb, and thus one would think that *mansa* is also from Spanish as well, but all indications point to the fact that it is not. *Mansa* means tame, peaceful, calm in Spanish, and the plant has no sedative effect, nor did local people ever use it as a calming agent. Its primary use is as an antimicrobial, antibacterial, and antifungal. The most likely explanation is that mansa is a Spanish alteration of the original native word for the plant, now lost in the depths of time."Hartweg, who collected it at León, Guanajuato in 1837, recorded the local name as *yerba del manso*. It is also known as *yerba del manso* in northern Baja California. The word "manso" could be short for "remanso" which would agree with the areas where the plant thrives.

Uses

Medicinal

Yerba mansa is used as an antimicrobial, an antibacterial, and to treat vaginal candidiasis. Yerba mansa is used to treat inflammation of the mucous membranes, swollen gums and sore throat. An infusion of roots can be taken as a diuretic to treat rheumatic diseases like gout by ridding the body of excess uric acid, which causes painful inflammation of the joints. Yerba mansa prevents the buildup of uric acid crystals in the kidneys which could cause kidney stones if left untreated. A powder of dried root can be sprinkled on infected areas to alleviate athlete's foot or diaper rash.

Yerba mansa is versatile, it can be taken orally as a tea, tincture, infusion or dried in capsule form. It can be used externally for soaking inflamed or infected areas. It can be ground and used as a dusting powder. Some people in Las Cruces, New Mexico use the leaves to make a poultice to relieve muscle swelling and inflammation.

Crafts

- Dried floral structures are used in dried arrangements.
- Dried plant parts (leaves, floral structure) emit a spicy fragrance and are used in potpourri.

Horticulture

- In the deserts of California, yerba mansa is being used as turf in public parks and ground cover in gardens.

ATRIPLEX

Atriplex is a plant genus of 250–300 species, known by the common names of saltbush and orache (or orach). It belongs to the subfamily Chenopodioideae of the family Amaranthaceae (which include the Chenopodiaceae of the Cronquist system). The genus is quite variable and widely distributed. It includes many desert and seashore plants and halophytes, as well as plants of moist environments. The generic name originated in Latin and was applied by Pliny the Elder to the edible oraches. The name saltbush derives from the fact that the plants retain salt in their leaves; they are able to grow in areas affected by soil salination.

Description

Atriplex patula, female flower with bracteoles and seed

The species in genus *Atriplex* are annual or perennial herbs, subshrubs, or shrubs. The plants are often covered with bladderlike hairs, that later collapse and form a silvery, scurfy or mealy surface, rarely with elongate trichomes. The alternate or rarely opposite leaves are petiolate or sessile, often persistent or tardily deciduous. The flat or slightly fleshy leaf blades are either entire, or serrate, or lobed and very variable in shape.

The inflorescences consist of axillary or terminal spikes or spicate panicles, or axillary clusters of glomeruled flowers. The flowers are unisexual, some species are monoecious, others dioecious. Male flowers consist of 3-5 perianth lobes and 3-5 stamens. Female flowers are usually lacking a perianth, but are enclosed by 2 foliaceous bracteoles, and contain an ovary with a short style and 2 stigmas.

In fruit, the bracteoles can enlarge, thicken or become appendaged. They enclose the fruit tightly, without becoming connate to it. The pericarp is adnate to the vertically orientated, flattened seed. The seed coat is thick, leathery or hardening. The annular embryo surrounds the perisperm.

The chromosome base number is $x = 9$, except for *Atriplex lanfrancoi*, which is $x=10$.

A few *Atriplex* species are C_3-plants, but most species are C_4-plants, with a characteristic leaf anatomy (kranz anatomy).

Distribution and Evolution

The genus *Atriplex* is distributed nearly worldwide from subtropical to temperate and to subarctic regions. Most species-rich are Australia, North America, South America and Eurasia. Many species are halophytes and are adapted to dry environments with salty soils.

The genus evolved in Middle Miocene, the C_4-photosynthesis pathway developed about 14.1-10.9 million years ago, when the climate became increasingly dry. The genus diversified rapidly and spread over the continents. The C_4*Atriplex* colonized North America probably from Eurasia during the Middle/Late Miocene, about 9.8-8.8 million years ago, and later spread to South America. Australia was colonized twice by two C_4lineages, one from Eurasia or America about 9.8-7.8 million years ago, and one from Central Asia about 6.3-4.8 million years ago. The last lineage diversified rapidly, and became the ancestor of most Australian *Atriplex* species.

Ecology

Atriplex species are used as food plants by the larvae of some Lepidoptera species; see the list of Lepidoptera which feed on *Atriplex*. For spiders such as *Phidippus californicus* and other arthropods, saltbush plants offer opportunities to hide and hunt in habitat that is otherwise often quite barren.

Use by Humans

The favored species for human consumption is now usually Garden Orache (*A. hortensis*), but many species are edible and the use of *Atriplex* as food is known since at least the late Epipaleolithic (Mesolithic).

Common Orache (*A. patula*) is attested as an archaeophyte in northern Europe, and the Ertebølle culture is presumed to have used it as a food. Its seed has been found among apparent evidence of cereal preparation and cooking at Late Iron Age villages in Britain. In the biblical Book of Job, *mallûah* is mentioned as food eaten by social outcasts (Job 30:4). Grey Saltbush (*A. cinerea*) is used as bushfood in Australia since prehistoric times. Chamiso (*A. canescens*) and Shadscale (*A. confertifolia*) were eaten by Native Americans, and Spearscale (*A. hastata*) was a food in rural Eurasia.

Meat from sheep which have grazed on saltbush has surprisingly high levels of vitamin E, is leaner and more hydrated than regular lamb and has consumer appeal equal to grain-fed lamb. The vitamin E levels could have animal health benefits while extending the shelf-life and maintaining the fresh red colour of saltbush lamb. This effect has been demonstrated for Old Man Saltbush (*A. nummularia*) and River Saltbush (*A. amnicola*). For reasons unknown, sheep seem to prefer the more fibrous, less nutritious River Saltbush.

Saltbushes are also used as an ornamental plant in landscaping and can be used to prevent soil erosion in coastal areas. Old Man Saltbush has also been successfully used to rehabilitate old mining sites around Lightning Ridge (Australia).

Atriplex elegans

Desert Holly (*Atriplex hymenelytra*) leaves and bracts

Atriplex latifolia

Silvery Saltbush *Atriplex lentiformis*

Old Man SaltbushAtriplex nummularia

Atriplex suberecta

Systematics

The genus *Atriplex* was first described in 1753 by Carl Linnaeus. The type species (lectotype) is *Atriplex hortensis*.

Atriplex is an extremely species rich genus and comprises about 250-300 species. After phylogenetical research, Kadereit et al. (2010) excluded *Halimione* as a distinct sister genus. The remaining *Atripex* species were grouped into several clades, that mostly are not identical with the traditional sections:

- ***Atriplex lanfrancoi/cana*-Clade**:
  - *Atriplex lanfrancoi* (Brullo & Pavone) G. Kadereit et Sukhor. (Syn.: *Cremnophyton lanfrancoi* Brullo & Pavone): endemic in Malta and Gozo.
  - *Atriplex cana* C.A. Mey.: from Eastern European Russia to Western China
- ***Atriplex* section *Atriplex***: annual C_3-plants.
  - *Atriplex aucheri* Moq.: in Eastern Europe and West Asia
  - *Atriplex hortensis* L. – Garden Orache, Red Orach, Mountain Spinach, French Spinach: in Asia, cultivated or naturalized in Europe
  - *Atriplex oblongifolia* Waldst. & Kit. – Oblong-leaved Orache: in Eurasia
  - *Atriplex sagittata* Borkh. (Syn.: *Atriplex nitens* Schkuhr): in Eurasia

- ***Atriplex* section *Teutliopsis*** Dumort.: annual C_3-plants
  - *Atriplex australasica* Moq.
  - *Atriplex calotheca* (Rafn) Fr.: in Northern Europe
  - *Atriplex crassifolia* Ledeb.: in Asia
  - *Atriplex davisii* Aellen: from Southern Europe to Egypt
  - *Atriplex glabriuscula* Edmondston – Northeastern Saltbush, Babington's Orache, Smooth Orache, Scotland Orache, Glabrous Orache: In Central and Northern Europe
  - *Atriplex gmelinii* C.A. Mey. ex Bong. – Gmelin's Saltbush: in Asia and North America
  - *Atriplex intracontinentalis* Sukhor.: from Central Europe to Asia
  - *Atriplex laevis* C.A. Mey.: in Asia, naturalized in Eastern Europe
  - *Atriplex latifolia* Wahlenb.: in Eurasia
  - *Atriplex littoralis* L. – Grass-leaved Orache: in Eurasia and North Africa
  - *Atriplex longipes* Drejer – Long-stalked Orache: in Northern Europe
  - *Atriplex micrantha* C.A. Mey.: in Asia, naturalized in Europe
  - *Atriplex nudicaulis* Boguslaw – Baltic Saltbush: in Eurasia
  - *Atriplex patula* L. – Common Orache, Spreading Orache: in Eurasia and North Africa
  - *Atriplex praecox* Hülph. – Early Orache>: in Northern Europe
  - *Atriplex prostrata* Moq. – Spear-leaved Orache, Thin-leaved Orache, Triangle Orache, fat hen: in Eurasia and North Africa
- **C_4-*Atriplex*-Clade**: containing the majority of species. The traditional classification into sections (sect. *Obione*, sect. *Pterochiton*, sect. *Psammophila*, sect. *Sclerocalymma*, sect.*Stylosa*) did not reflect the phylogenetical relationships and was rejected by Kadereit et al. (2010).
  - *Atriplex acanthocarpa* (Torr.) S. Watson: in North America
  - *Atriplex acutibractea* Anderson: in Australia
  - *Atriplex altaica* Sukhor.: in Asia
  - *Atriplex angulata* Benth.: in Australia
  - *Atriplex billardierei* (Moq.) Hook. f.: in Australia
  - *Atriplex canescens* (Pursh) Nutt. – Chamiso, Chamiza, Four-winged Saltbush, Grey Sagebrush: in North America
  - *Atriplex centralasiatica* Iljin: in Asia
  - *Atriplex cinerea* Poir. – Grey Saltbush, Truganini: in Australia
  - *Atriplex codonocarpa* P.G. Wilson: in Australia
  - *Atriplex conduplicata* F. Muell.: in Australia

- *Atriplex confertifolia* (Torr. & Frém.) S. Watson – Shadscale (Saltbush): in North America
- *Atriplex cordobensis* Gand. & Stuck.: in South America
- *Atriplex deserticola* Phil.: in South America
- *Atriplex dimorphostegia* Kar. & Kir.: in North Africa
- *Atriplex eardleyae* Aellen: in Australia
- *Atriplex elachophylla* F. Muell.: in Australia
- *Atriplex fissivalvis* F. Muell.: in Australia
- *Atriplex flabellum* Bunge ex Boiss.: in Eurasia
- *Atriplex gardneri* (Moq.) D. Dietr. – Gardner's saltbush, Moundscale: in North America
- *Atriplex glauca* L.: in Portugal, Spain and in North Africa
- *Atriplex halimus* L. – Mediterranean Saltbush, Sea Orache, Shrubby Orache: in South Europe, North Africa and Southwest Asia
- *Atriplex herzogii* Standl.: in North America
- *Atriplex holocarpa* F. Muell.: in Australia
- *Atriplex hymenelytra* (Torr.) S. Watson – Desert Holly: in North America
- *Atriplex hymenotheca* Moq.: in Australia
- *Atriplex imbricata* (Moq.) D. Dietr.: in South America
- *Atriplex inamoena* Aellen: in Eurasia
- *Atriplex intermedia* Anderson: in Australia
- *Atriplex isatidea* Moq.: in Australia
- *Atriplex laciniata* L. – Frosted Orache: In Western and Northern Europe
- *Atriplex lampa* (Moq.) Gillies ex Small: in South America
- *Atriplex lehmanniana* Bunge: in Eurasia
- *Atriplex lentiformis* (Torr.) S. Watson – Quail Bush: in North America
- *Atriplex leptocarpa* F. Muell.: in Australia
- *Atriplex leucoclada* Boiss.: in Eurasia
- *Atriplex leucophylla* (Moq.) D. Dietr.: in North America
- *Atriplex lindleyi* Moq.: in Australia
- *Atriplex moneta* Bunge ex Boiss.: in Eurasia
- *Atriplex muelleri* Benth.: in Australia
- *Atriplex nessorhina* S.W.L. Jacobs: in Australia
- *Atriplex nummularia* Lindl. – Old Man Saltbush, Giant Saltbush: in Australia
- *Atriplex obovata* Moq.: in North America
- *Atriplex pamirica* Iljin: in Eurasia
- *Atriplex parishii* S. Watson: in North America

- *Atriplex parryi* S. Watson: in North America
- *Atriplex parvifolia* Kunth: in South America
- *Atriplex patagonica* (Moq.) D. Dietr.: in South America
- *Atriplex phyllostegia* (Torr. ex S. Watson) S. Watson: in North America
- *Atriplex polycarpa* (Torr.) S. Watson – Allscale (Saltbush), Desert Saltbush, Cattle Saltbush, Cattle Spinach: in North America
- *Atriplex powellii* S. Watson – Powell's Saltbush: in North America
- *Atriplex pseudocampanulata* Aellen: in Australia
- *Atriplex quinii* F. Muell.: in Australia
- *Atriplex recurva* d'Urv.: in Eurasia, endemic to areas around the Aegean
- *Atriplex rhagodioides* F. Muell.: in Australia
- *Atriplex rosea* L. – Tumbling Orache: in Eurasia and North Africa
- *Atriplex rusbyi* Britton ex Rusby: in South America
- *Atriplex schugnanica* Iljin: in Asia
- *Atriplex semibaccata* R. Br. – Australian Saltbush, Berry Saltbush, Creeping Saltbush: in Australia
- *Atriplex semilunaris* Aellen: in Australia
- *Atriplex serenana* A. Nelson ex Abrams: in North America
- *Atriplex sibirica* L.; in Asia, naturalized in Europe
- *Atriplex sphaeromorpha* Iljin: in Russia, Ukraine and Caucasus
- *Atriplex spinibractea* Anderson: in Australia
- *Atriplex spongiosa* F. Muell.: in Australia
- *Atriplex stipitata* Benth.: in Australia
- *Atriplex sturtii* S.W.L. Jacobs: in Australia
- *Atriplex suberecta* I. Verd. – sprawling saltbush, lagoon saltbush: in Australia
- *Atriplex tatarica* Aellen: in Europe, North Africa and Asia
- *Atriplex turbinata* (Anderson) Aellen: in Australia
- *Atriplex undulata* (Moq.) D. Dietr.: in South America
- *Atriplex velutinella* F. Muell.: in Australia
- *Atriplex vesicaria* Heward ex Benth. – Bladder Saltbush: in Australia

Selected species not yet investigated phylogenetically:

- *Atriplex acadiensis* – Maritime Saltbush
- *Atriplex acutiloba*
- *Atriplex alaskensis* S. Watson – Alaska Orache
- *Atriplex amnicola* Paul G. Wilson – River Saltbush, Swamp Saltbush: in Australia, naturalized in California

- *Atriplex × aptera*
- *Atriplex argentea* – Silvery Saltbush
- *Atriplex asterocarpa*
- *Atriplex barclayana*
- *Atriplex bonnevillensis*
- *Atriplex bunburyana* F.Muell.
- *Atriplex cephalantha*
- *Atriplex cordifolia*
- *Atriplex cordulata*
- *Atriplex cornigera*
- *Atriplex coronata* S. Watson: in California
- *Atriplex corrugata*
- *Atriplex coulteri*
- *Atriplex crassipes*
- *Atriplex cristata*
- *Atriplex cryptocarpa*
- *Atriplex cuneata*
- *Atriplex depressa*
- *Atriplex dioica* – Thick-leaved Orache, Saline Saltbush
- *Atriplex drymarioides*
- *Atriplex eichleri*
- *Atriplex elegans*
- *Atriplex erecticaulis*
- *Atriplex exilifolia* F.Muell.
- *Atriplex falcata* – Sickle Saltbush
- *Atriplex flabelliformis*
- *Atriplex franktonii* – Frankton's Saltbush
- *Atriplex fruticulosa*
- *Atriplex garrettii*
- *Atriplex graciliflora*
- *Atriplex griffithsii* – Griffith's Saltbush (sometimes included in *A. lentiformis* or *A. torreyi*)
- *Atriplex hastata* – Halberd-leaved Orache, Spearscale (often included in *A. patula*)
- *Atriplex heterosperma* Bunge – Two-scaled Orache: in Eastern Europe
- *Atriplex humifusa*
- *Atriplex humilis* F.Muell.

- *Atriplex hypoleuca*
- *Atriplex incrassata* F.Muell.
- Atriplex infrequens
- Atriplex johnsonii
- Atriplex johnstonii
- *Atriplex klebergorum*
- *Atriplex kochiana* Maiden
- *Atriplex limbata*
- *Atriplex lobativalvis* F.Muell.
- *Atriplex macropterocarpa*
- *Atriplex matamorensis*
- *Atriplex maximowicziana* – Maximowicz's Saltbush
- *Atriplex minuscula*
- *Atriplex minuticarpa*
- *Atriplex morrisii*
- *Atriplex nana*
- *Atriplex navajoensis*
- *Atriplex nuttallii* S. Watson – Nuttall's Saltbush: in North America
- *Atriplex obconica*
- *Atriplex pachypoda*
- *Atriplex pacifica*
- *Atriplex paludosa* R.Br.
- *Atriplex papillata*
- *Atriplex persistens*
- *Atriplex plebeja*
- *Atriplex pumilio* R.Br.
- *Atriplex pusilla*
- *Atriplex quadrivalvata*
- *Atriplex saccaria*
- *Atriplex spinifera*
- *Atriplex spinulosa*
- *Atriplex subspicata*
- *Atriplex subtilis*
- *Atriplex tenuissima*
- *Atriplex texana*
- *Atriplex torreyi*

- *Atriplex tridentata*
- *Atriplex truncata* – Wedge-leaved Saltbush
- *Atriplex tularensis*
- *Atriplex valdesii* Flores Olv., Mexico
- *Atriplex vallicola*
- *Atriplex vestita*
- *Atriplex wardii*
- *Atriplex watsonii*
- *Atriplex welshii*
- *Atriplex wolfii*
- *Atriplex wrightii*

Excluded species:

- *Extriplex*:
    - *Extriplex californica* (Moq.) E.H.Zacharias (Syn. Atriplex californica Moq.)
    - *Extriplex joaquiniana* (A.Nelson) E.H.Zacharias (Syn. Atriplex joaquiniana A.Nelson)
- *Halimione Aellen:*
    - *Halimione pedunculata* (L.) Aellen (Syn.: Atriplex pedunculata L.)
    - *Halimione portulacoides* (L.) Aellen (Syn.: Atriplex portulacoides L.)
    - *Halimione verrucifera* (M.Bieb.) Aellen (Syn.: Atriplex verrucifera M.Bieb.)
- *Stutzia E.H.Zacharias:*
    - *Stutzia covillei* (Standl.) E. H. Zacharias (Syn. Atriplex covillei (Standl.) J. F. Macbr.)
    - *Stutzia dioica* (Nutt.) E. H. Zacharias (Syn. Atriplex suckleyi (Torrey) Rydberg, Atriplex ovata (Rydb.) Clem. & E. G. Clem.)

ATTALEA SPECIOSA

Attalea speciosa (babassu, babassu palm, babaçu, cusi) is a palm native to the Amazon Rainforest region in South America. The babassu palm is the predominant species in the Maranhão Babaçu forests of Maranhão and Piauí states.

Babassu palms in Brazil

This plant has commercial value because its seeds produce an edible oil called babassu oil, which is also used in cleaners and skin care products. The fruit is used to produce products such as medicines, beauty aids, and beverages. Traditional communities of the Maranhão region also produce a flour from the fruit and this is commercialized as a nutritional supplement. The leaves are also used to provide thatch for houses and can be woven into mats for constructing house walls. The stems are used for timbers. The Babassu palm is considered a weed in pasture areas of Cerradovegetation in Brazil.

Its fruits and leaves are one of the main materials used by the artisans of Tocantins (Brazil) to manufacture different products very popular with tourists.

Taxonomy

In his 1995 *The Palms of the Amazon*, Andrew Henderson recognised *A. speciosa* and *A. spectabilis* as valid species, but considered the latter to either be an acaulescent form of *A. speciosa* or a hybrid between it and *A. microcarpa*. In their 2005 *World Checklist of Palms* Rafäel Govaerts and John Dransfield accepted *A. spectabilis*as a valid taxon but Sidney F Glassman considered it a dubious taxon in his 1999 *Taxonomic Treatment of Palm Subtribe Attaleinae*. *Attalea vitrivir* was recognised as a distinct species by Michael Balick and coauthors; Glassman and Govaerts and Dransfield concurred, but Henderson considered it part of *A. speciosa*. Glassman also described a fourth member of this group, *A. brejinhoensis*, and it is accepted by Govaerts and Dransfield.

Reproduction and Growth

Attalea speciosa is monoecious—male and female flowers are separate, but are borne by the same plant. Pollination has been attributed both to insects and wind.

Uses

Biofuel

In February 2008, babassu palm oil was used in a blend with coconut oil and jet fuel to power one engine of a Virgin Atlantic Boeing 747 during a flight test. It is considered to be less environmentally friendly than biofuel made from the *Jatropha* plant that Air New Zealand tested in December 2008. This is because Babassu grows in forested areas that require deforestation, rather than in arid land and land with marginal use, like the Jatropha plant.

Other

Babassu oil can also be used in food cooking, as a lubricant and also in soap and cosmetics. The shell of the nut can be used to make smoke-less charcoal and flesh of the babassu nut to produce flour. Palm hearts are extracted from the tree for to make a juice that used as medicine for its antiseptic and styptic properties. The leaf and stalk of the babassu palm is used in building materials. Baskets and other handicrafts can be made from the fibre of the leaves.

PANICUM VIRGATUM

Panicum virgatum, commonly known as switchgrass, is a perennial warm season bunchgrass native to North America, where it occurs naturally from 55°N latitude in Canada southwards into the United States and Mexico. Switchgrass is one of the dominant species of the central North American tallgrass prairie and can be found in remnant prairies, in native grass pastures, and naturalized along roadsides. It is used primarily for soil conservation, forage production, game cover, and as an ornamental grass. More recently, it has also been used as a biomass crop for ethanol and butanol, in phytoremediation projects, fiber, electricity, and heat production, and for biosequestration of atmospheric carbon dioxide.

Other common names for switchgrass include tall panic grass, Wobsqua grass, blackbent, tall prairiegrass, wild redtop, thatchgrass, and Virginia switchgrass.

Description

Root system of switchgrass grown at the Land Institute

Switchgrass is a hardy, deep-rooted, perennialrhizomatous grass that begins growth in late spring. It can grow up to 2.7 m high, but is typically shorter than big bluestem grass or indiangrass. The leaves are 30–90 cm long, with a prominent midrib. Switchgrass uses C_4 carbon fixation, giving it an advantage in conditions of drought and high temperature. Its flowers have a well-developed panicle, often up to 60 cm long, and it bears a good crop of seeds. The seeds are 3–6 mm long and up to 1.5 mm wide, and are developed from a single-flowered spikelet. Both glumes are present and well developed. When ripe, the seeds sometimes take on a pink or dull-purple tinge, and turn golden brown with the foliage of the plant in the fall. Switchgrass is both a perennial and self-seeding crop, which means farmers do not have to plant and reseed after annual harvesting. Once established, a switchgrass stand can survive for ten years or longer. Unlike corn, switchgrass can grow on marginal lands and requires relatively modest levels of chemical fertilizers. Overall, it is considered a resource-efficient, low-input crop for producing bioenergy from farmland.

Habitat

Much of North America, especially the prairies of the Midwestern United States, was once prime habitat to vast swaths of native grasses, including switchgrass, indiangrass (*Sorghastrum nutans*), eastern gamagrass (*Tripsacum dactyloides*), big bluestem (*Andropogon gerardi*), little bluestem (*Schizachyrium scoparium*), bluestem prairie (*Andropogon panicum sorghastrum*) and others. As European settlers began spreading west across the continent, the native grasses were plowed under and the land converted to crops such as corn, wheat, and oats. Introduced grasses such as fescue, bluegrass, and orchardgrass also replaced the native grasses for use as hay and pasture for cattle.

Distribution

Switchgrass is a versatile and adaptable plant. It can grow and even thrive in many weather conditions, lengths of growing seasons, soil types, and land conditions. Its distribution spans south of latitude 55°N from Saskatchewan to Nova Scotia, south over most of the United States east of the Rocky Mountains, and further south into Mexico. As a warm-season perennial grass, most of its growth occurs from late spring through early fall; it becomes dormant and unproductive during colder months. Thus, the productive season in its northern habitat can be as short as three months, but in the southern reaches of its habitat the growing season may be as long as eight months, around the Gulf Coast area.

Switchgrass is a diverse species, with striking differences between plants. This diversity, which presumably reflects evolution and adaptation to new environments as the species spread across the continent, provides a range of valuable traits for breeding programs. Switchgrass has two distinct forms, or "cytotypes": the lowland cultivars, which tend to produce more biomass, and the upland cultivars, which are generally of more northern origin, more cold-tolerant, and therefore usually preferred in northern areas. Upland switchgrass types are generally shorter ($\leq$ 2.4 m, tall) and less coarse than lowland types. Lowland cultivars may grow to $\geq$ 2.7 m, in favorable environments. Both upland and lowland cultivars are deeply rooted (> 1.8 m, in favorable soils) and have short rhizomes. The upland types tend to have more vigorous rhizomes, so the lowland cultivars may appear to have a bunchgrass habit, while the upland types tend to be more sod-forming. Lowland cultivars appear more plastic in their morphology, produce larger plants if stands become thin or when planted in wide rows, and they seem to be more sensitive to moisture stress than upland cultivars.

In native prairies, switchgrass is historically found in association with several other important native tallgrass prairie plants, such as big bluestem, indiangrass, little bluestem, sideoats grama, eastern gamagrass, and various forbs (sunflowers, gayfeather, prairie clover, and prairie coneflower). These widely adapted tallgrass species once occupied millions of hectares.

Switchgrass' suitability for cultivation in the Gran Chaco is being studied by Argentina's Instituto Nacional de Tecnología Agropecuaria (INTA).

Establishment and Management

Switchgrass can be grown on land considered unsuitable for row crop production, including land that is too erodible for corn production, as well as sandy and gravelly soils in humid regions that typically produce low yields of other farm crops. No single method of establishing switchgrass can be suggested for all situations. The crop can be established both by no-till and conventional tillage. When seeded as part of a diverse mixture, planting guidelines for warm-season grass mixtures for conservation plantings should be followed. Regional guidelines for growing and managing switchgrass for bioenergy or conservation plantings are available. Several key factors can increase the likelihood of success for establishing switchgrass. These include:

- Planting switchgrass after the soil is well warmed during the spring.
- Using seeds that are highly germinable and planting 0.6 - 1.2 cm deep, or up to 2 cm deep in sandy soils.
- Packing or firming the soil both before and after seeding.
- Providing no fertilization at planting to minimize competition.
- Controlling weeds with chemical and/or cultural control methods.

Mowing and properly labeled herbicides are recommended for weed control. Chemical weed control can be used in the fall prior to establishment, or before or after planting. Weeds should be mowed just above the height of the growing switchgrass. Hormone herbicides, such as 2,4-D, should be avoided as they are known to reduce development of switchgrass when applied early in the establishing year. Plantings that appear to have failed due to weed infestations are often wrongly assessed, as the failure is often more apparent than real. Switchgrass stands that are initially weedy commonly become well established with appropriate management in subsequent years. Once established, switchgrass can take up to three years to reach its full production potential. Depending on the region, it can typically produce 1/4 to 1/3 of its yield potential in its first year and 2/3 of its potential in the year after seeding.

After establishment, switchgrass management will depend on the goal of the seeding. Historically, most switchgrass seedings have been managed for the Conservation Reserve Program in the US. Disturbance such as periodic mowing, burning, or disking is required to optimize the stand's utility for encouraging biodiversity. Increased attention is being placed on switchgrass management as an energy crop. Generally, the crop requires modest application of nitrogen fertilizer, as it is not a heavy feeder. Typical nitrogen (N) content of senescent material in the fall is 0.5% N. Fertilizer nitrogen applications of about 5 kg N/hectare (ha) applied for each tonne of biomass removed is a general guideline. More specific recommendations for fertilization are available regionally in North America. Herbicides are not often used on switchgrass after the seeding year, as the crop is generally quite competitive with weeds. Most bioenergy conversion processes for switchgrass, including those for cellulosic ethanol and pellet fuel production, can generally accept some alternative species in the harvested biomass. Stands of switchgrass should be harvested no more than twice per year, and one cutting often provides as much biomass as two. Switchgrass can be harvested with the same field equipment used for hay production, and it is well-suited to baling or bulk field harvesting. If its biology is properly taken into consideration, switchgrass can offer great potential as an energy crop.

Uses

Switchgrass can be used as a feedstock for biomass energy production, forages, and grazing, as ground cover for soil conservation, and as a substrate for growing mushrooms. It can also be used to control erosion. It can be used by cattle farmers for hay and pasture and as a substitute for wheat straw in many applications, including livestock bedding, straw bale housing.

Panicum virgatum 'Heavy Metal', an ornamental switchgrass, in early summer

Additionally, switchgrass is grown as a drought-resistant ornamental grass in average to wet soils and in full sun to part shade.

Bioenergy

Switchgrass has been researched as a renewable bioenergy crop since the mid-1980s, because it is a native perennial warm season grass with the ability to produce moderate to high yields on marginal farmlands. It is now being considered for use in several bioenergy conversion processes, including cellulosic ethanol production, biogas, and direct combustion for thermal energy applications. The main agronomic advantages of switchgrass as a bioenergy crop are its stand longevity, drought and flooding tolerance, relatively low herbicide and fertilizer input requirements, ease of management, hardiness in poor soil and climate conditions, and widespread adaptability in temperate climates. In some warm humid southern zones, such as Alabama, it has the ability to produce up to 25 oven-dry tonnes per hectare (ODT/ha). A summary of switchgrass yields across 13 research trial sites in the United States found the top two cultivars in each trial to yield 9.4 to 22.9 t/ha, with an average yield of 14.6 ODT/ha. However, these yields were recorded on small plot trials, and commercial field sites could be expected to be at least 20% lower than these results. In the United States, switchgrass yields appear to be highest in warm humid regions with long growing seasons such as the US Southeast and lowest in the dry short season areas of the Northern Great Plains. The energy inputs required to grow switchgrass are favorable when compared with annual seed bearing crops such as corn, soybean, or canola, which can require relatively high energy inputs for field operations, crop drying, and fertilization. Whole plant herbaceous perennial C4 grass feedstocks are desirable biomass energy feedstocks, as they require fewer fossil energy inputs to grow and effectively capture solar energy because of their C4 photosynthetic system and perennial nature. One study cites it takes from 0.97 to 1.34 GJ to produce 1 tonne of switchgrass, compared with 1.99 to 2.66 GJ to produce 1 tonne of corn. Another study found that switchgrass uses 0.8 GJ/ODT of fossil energy compared to grain corn's 2.9 GJ/ODT. Given that switchgrass contains approximately 18.8 GJ/ODT of biomass, the energy output-to-input ratio for the crop can be up to 20:1. This highly favorable ratio is attributable to its relatively high energy output per hectare and low energy inputs for production.

Considerable effort is being expended in developing switchgrass as a cellulosic ethanol crop in the USA. In George W. Bush's 2006 State of the Union Address, he proposed using switchgrass for ethanol; since then, over US$100 million has been invested into researching switchgrass as a potential biofuel source. Switchgrass has the potential to produce up to 380 liters of ethanol per tonne harvested. However, current technology for herbaceous biomass conversion to ethanol is about 340 liters per tonne. In contrast, corn ethanol yields about 400 liters per tonne. There are substantial efforts being made to increase the yield of corn derived ethanol:

> (Corn-) Ethanol yield has already improved from 2.4 gallons per bushel in the 1980s to 2.8 gallons in modern plants. Corn hybrids developed specifically for ethanol production have demonstrated ethanol yield increases of 2.7 percent—and using the cellulose (fiber) in the corn kernel, in addition to the starch, could increase yield by another 10 to 13 percent. With this combination of hybrid and process optimization, theoretical yields of 3.51 gallons of ethanol per bushel are within reason—with no negative impact on protein or oil content for animal feed uses of the distillers grains.

Process improvements for the conventional corn-based industry are based on new technologies such as cavitation, new catalysis and new enzymes.

The main advantage of using switchgrass over corn as an ethanol feedstock is its cost of production is generally about 1/2 that of grain corn, and more biomass energy per hectare can be captured in the field. Thus, switchgrass cellulosic ethanol should give a higher yield of ethanol per hectare at lower cost. However, this will depend on whether the cost of constructing and operating cellulosic ethanol plants can be reduced considerably. The switchgrass ethanol industry energy balance is also considered to be substantially better than that of corn ethanol. During the bioconversion process, the lignin fraction of switchgrass can be burned to provide sufficient steam and electricity to operate the biorefinery. Studies have found that for every unit of energy input needed to create a biofuel from switchgrass, four units of energy are yielded. In contrast, corn ethanol yields about 1.28 units of energy per unit of energy input. A recent study from the Great Plains indicated that for ethanol production from switchgrass, this figure is 6.4, or alternatively, that 540% more energy was contained in the ethanol produced than was used in growing the switchgrass and converting it to liquid fuel. However, there remain commercialization barriers to the development of cellulosic ethanol technology. Projections in the early 1990s for commercialization of cellulosic ethanol by the year 2000 have not been met. The commercialization of cellulosic ethanol is thus proving to be a significant challenge, despite noteworthy research efforts.

Thermal energy applications for switchgrass appear to be closer to near-term scale-up than cellulosic ethanol for industrial or small-scale applications. For example, switchgrass can be pressed into fuel pellets that are subsequently burned in pellet stoves used to heat homes (which typically burn corn or wood pellets). Switchgrass has been widely tested as a substitute for coal in power generation. The most widely studied project to date has been the Chariton Valley Project in Iowa. The Show-Me-Energy Cooperative (SMEC) in Missouri is using switchgrass and other warm-season grasses, along with wood residues, as feedstocks for pellets used for the firing of a coal-fired power plant. In Eastern Canada, switchgrass is being used on a pilot scale as a feedstock for commercial heating applications. Combustion studies have been undertaken and it appears to be well-suited as a commercial boiler fuel. Research is also being undertaken to develop switchgrass as a pellet fuel because of lack of surplus wood residues in eastern Canada, as a slowdown in the forest products industry in 2009 is now resulting in wood pellet shortages throughout Eastern North America. Generally speaking, the direct firing of switchgrass for thermal applications can provide the highest net energy gain and energy output-to-input ratio of all switchgrass bioconversion processes. Research has found switchgrass, when pelletized and used as a solid biofuel, is a good candidate for displacing fossil fuels. Switchgrass pellets were identified to have a 14.6:1 energy output-to-input ratio, which is substantially better than that for liquid biofuel options from farmland. As a greenhouse gas mitigation strategy, switchgrass pellets were found to be an effective means to use farmland to mitigate greenhouse gases on the order of 7.6-13 tonnes per hectare of CO_2. In contrast, switchgrass cellulosic ethanol and corn ethanol were found to mitigate 5.2 and 1.5 tonnes of CO_2 per hectare, respectively.

Historically, the major constraint to the development of grasses for thermal energy applications has been the difficulty associated with burning grasses in conventional boilers, as biomass quality problems can be of particular concern in combustion applications. These technical problems now appear to have been largely resolved through crop management practices such as fall mowing and spring harvesting that allow for leaching to occur, which leads to fewer aerosol-forming compounds (such as K and Cl) and N in the grass. This reduces clinker formation and corrosion, and enables switchgrass to be a clean combustion fuel source for use in smaller combustion appliances. Fall harvested grasses likely have more application for larger commercial and industrial boilers. Switchgrass is also being used to heat small industrial and farm buildings in Germany and China through a process used to make a low quality natural gas substitute.

Bai et al. (2010) conducted a study to analyze the environmental sustainability of using switchgrass plant material as a feedstock for ethanol production. Life cycle analysis was used to make this assessment. They compared efficiency of E10, E85, and ethanol with gasoline. They took into account air and water emissions associated with growing, managing, processing and storing the switchgrass crop. They also factored in the transportation of the stored switchgrass to the ethanol plant where they assumed the distance was 20 km. The reductions in global warming potential by using E10 and E85 were 5 and 65%, respectively. Their models also suggested that the "human toxicity potential" and "eco-toxicity potential" were substantially greater for the high ethanol fuels (i.e., E85 and ethanol) than for gasoline and E10.

In 2014, a genetically altered form of the bacterium Caldicellulosiruptor bescii, which can cheaply and efficiently turn switchgrass into ethanol, was created.

Biodegradable Plastics Production

In a novel application, US scientists have genetically modified switchgrass to enable it to produce polyhydroxybutyrate, which accumulates in beadlike granules within the plant's cells. In preliminary tests, the dry weight of a plants leaves were shown to comprise up to 3.7% of the polymer. Such low accumulation rates do not, as of 2009, allow for commercial use of switchgrass as a biosource.

Soil Conservation

Switchgrass is useful for soil conservation and amendment, particularly in the United States and Canada, where switchgrass is endemic. Switchgrass has a deep fibrous root system – nearly as deep as the plant is tall. Since it, along with other native grasses and forbs, once covered the plains of the United States that are now the Corn Belt, the effects of the past switchgrass habitat have been beneficial, lending to the fertile farmland that exists today. The deep fibrous root systems of switchgrass left a deep rich layer of organic matter in the soils of the Midwest, making those mollisol soils some of the most productive in the world. By returning switchgrass and other perennial prairie grasses as an agricultural crop, many marginal soils may benefit from increased levels of organic material, permeability, and fertility, due to the grass's deep root system.

Soil erosion, both from wind and water, is of great concern in regions where switchgrass grows. Due to its height, switchgrass can form an effective wind erosion barrier. Its root system, also, is excellent for holding soil in place, which helps prevent erosion from flooding and runoff. Some highway departments (for example, KDOT) have used switchgrass in their seed mixes when re-establishing growth along roadways. It can also be used on strip mine sites, dikes, and pond dams. Conservation districts in many parts of the United States use it to control erosion in grass waterways because of its ability to anchor soils while providing habitat for wildlife.

Forages and Grazing

Switchgrass is an excellent forage for cattle; however, it has shown toxicity in horses, sheep, and goats through chemical compounds known as saponins, which cause photosensitivity and liver damage in these animals. Researchers are continuing to learn more about the specific conditions under which switchgrass causes harm to these species, but until more is discovered, it is recommended switchgrass not be fed to them. For cattle, however, it can be fed as hay, or grazed.

Grazing switchgrass calls for watchful management practices to ensure survival of the stand. It is recommended that grazing begin when the plants are about 50 cm tall, and that grazing be discontinued when the plants have been eaten down to about 25 cm, and to rest the pasture 30 – 45 days between grazing periods. Switchgrass becomes stemmy and unpalatable as it matures, but during the target grazing period, it is a favorable forage with a relative feed value (RFV) of 90-104. The grass's upright growth pattern places its growing point off the soil surface onto its stem, so leaving 25 cm of stubble is important for regrowth. When harvesting switchgrass for hay, the first cutting occurs at the late boot stage – around mid-June. This should allow for a second cutting in mid-August, leaving enough regrowth to survive the winter.

Game Cover

Switchgrass is well-known among wildlife conservationists as good forage and habitat for upland game bird species, such as pheasant, quail, grouse, and wild turkey, and song birds, with its plentiful small seeds and tall cover. A study published in 2015 has shown that switchgrass, when grown in a traditional monoculture, has an adverse impact on some wildlife. Depending on how thickly switchgrass is planted, and what it is partnered with, it also offers excellent forage and cover for other wildlife across the country. Aesthetic switchgrass stands on farms of producers attracts an abundance of wildlife, which is considered an environmental benefit. Some members of Prairie Lands Bio-Products, Inc. in Iowa have even turned this benefit into a profitable business by leasing their switchgrass land for hunting during the proper seasons. The benefits to wildlife can be extended even in large-scale agriculture through the process of strip harvesting, as recommended by The Wildlife Society, which suggests that rather than harvesting an entire field at once, strip harvesting could be practiced so that the entire habitat is not removed, thereby protecting the wildlife inhabiting the switchgrass.

SALICORNIA

Salicornia is a genus of succulent, halophyte (salt tolerant) flowering plants in the family Amaranthaceae that grow in salt marshes, on beaches, and among mangroves. Salicornia species are native to North America, Europe, South Africa, and South Asia. Common names for the genus include glasswort, pickleweed, and marsh samphire; these common names are also used for some species not in Salicornia. The main European species is often eaten, called marsh samphire in Britain, and the main North American species is occasionally sold in grocery stores or appears on restaurant menus, usually as 'sea beans' or samphire greens or sea asparagus .

Description

The *Salicornia* species are small annual herbs. They grow prostrate to erect, their simple or branched stems are succulent, glabrous, and apparently jointed. Older stems may be somewhat woody basally. The opposite leaves are fleshy, glabrous, sessile, basally connate and decurrent and enclosing the stem (thus forming the joints). The leaf blades are reduced to small collar-like scales with narrow scarious margin.Many species are green, but their foliage turns red in autumn.

All stems are terminating in spike-like apparently jointed inflorescences. Each joint consists of two opposite minute bracts with an (1-) 3-flowered cyme tightly embedded in cavities of the main axis and partly hidden by the bracts. The flowers are arranged in a triangle, both lateral flowers beneath the central flower. The hermaphrodite flowers are more or less radially symmetric, with a perianth of three fleshy tepals connate nearly to the apex. There are 1-2 stamens and an ovary with two stigmas.

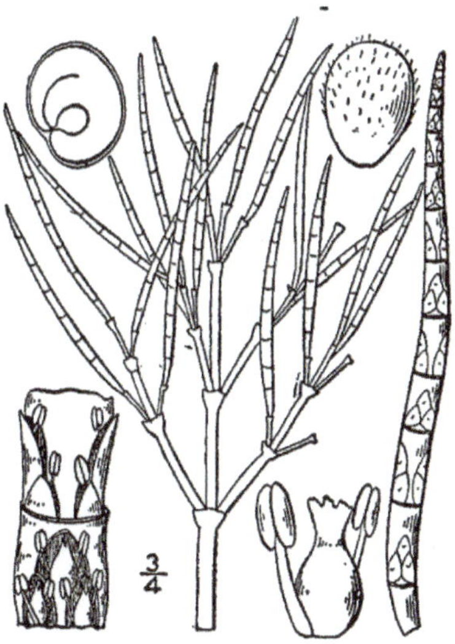

Illustration of *Salicornia maritima*

The perianth is persistent in fruit. The fruit wall (pericarp) is membranous. The vertical seed is ellipsoid, with yellowish brown, membranous, hairy seed coat. The seed contains no perisperm (feeding tissue).

Like most members of the subfamily Salicornioideae, *Salicornia* species use the C3 carbon fixation pathway to take in carbon dioxide from the surrounding atmosphere.

Distribution and Habitat

Salicornia rubra at a saline inland habitat

The species of *Salicornia* are widely distributed over the Northern Hemisphere and in southern Africa, ranging from the subtropics to subarctic regions. They are absent from South America and Australia.

They grow in coastal salt marshes and in inland salty habitats like shores of salt lakes. *Salicornia* species are halophytes and can generally tolerate immersion in salt water (hygrohalophytes).

Ecology

Salicornia species are used as food plants by the larvae of some Lepidoptera species, including the *Coleophora* case-bearers *C. atriplicis* and *C. salicorniae* (the latter feeds exclusively on *Salicornia* spp.).

Phylogenetics

The genus probably originated during the Miocene in the region between the Mediterranean basin and Central Asia. Evolving from within the perennial and frost-sensitive genus *Sarcocornia*, the annual, strongly inbreeding and frost-tolerant *Salicornia* diversified during the late Pliocene to early Pleistocene. By events of intercontinental dispersals, they reached southern Africa twice, North America at least three times. Two tetraploid lineages expanded rapidly, with the ability to colonize lower belts of the saltmarshes than their diploid relatives. Inbreeding and geographical isolation led to a large number of reproductive isolated species that are only weakly differentiated.

Systematics

Salicornia bigelovii

Salicornia depressa

The genus *Salicornia* was first described in 1753 by Carl Linnaeus. *Salicornia europaea* was selected as the type species.

The taxonomic classification of this genus is extremely difficult (and has been called a "taxonomic nightmare"), determination of species seems almost impossible for non-specialists. The reasons for those difficulties are the reduced habit with weak morphological differentiation, and high phenotypic variability. As the succulent plants lose their characteristics while drying, herbarium specimens often cannot be determined with certainty and are less suited for taxonomic studies.

Based on molecular genetic research (Kadereit et al. 2007, 2012), *Salicornia* comprises the following species:

In Eurasia:

- *Salicornia europaea* species group, Common glasswort, with two cryptic species:
    - *Salicornia europaea* L., with three subspecies
        - *Salicornia europaea* subsp. *europaea*, along coasts from southern Spain to northern Scandinavia
        - *Salicornia europaea* subsp. *disarticulata* (Moss) Lambinon & Vanderpoorten, Atlantic coasts of Brittany, the Netherlands and southern England.

    - *Salicornia europaea* subsp. × *marshallii* Lambinon & Vanderpoorten, the hybrid of both former subspecies, Atlantic coasts of Brittany and the Netherlands
  - *Salicornia perennans* Willd. (Syn. *Salicornia prostrata* Pallas), with two subspecies:
    - *Salicornia perennans* subsp. *perennans*, with wide mediterranean-continental distribution, from North Africa and the Mediterranean region to the Baltic Sea and White Sea (occurring locally at coasts of the Atlantic and North Sea), in Asia to Yakutsk (Siberia), Japan, and Korean Peninsula.
    - *Salicornia perennans* subsp. *altaica* (Lomon.) G. Kadereit & Piirainen, endemic in Altai Mountains (Russia, Mongolia)
- *Salicornia procumbens* species group:
  - *Salicornia procumbens* Sm., with four subspecies:
    - *Salicornia procumbens* subsp. *procumbens*, widely distributed along the Mediterranean and Atlantic coasts from Morocco to Scandinavia, inland occurrences in Turkey and Ukraine.
    - *Salicornia procumbens* subsp. *freitagii* (Yaprak & Yardakulol) G. Kadereit & Piirainen, endemic in Turkey (Anatolia).
    - *Salicornia procumbens* subsp. *pojarkovae* (Semenova) G. Kadereit & Piirainen, coasts of the White Sea (Russia) and Barents Sea (Norway).
    - *Salicornia procumbens* subsp. *heterantha* (S.S. Beer & Demina) G. Kadereit & Piirainen, endemic in southeast European Russia (Rostov Oblast).
  - *Salicornia persica* Akhani, with two subspecies:
    - *Salicornia persica* subsp. *persica*, in Iran
    - *Salicornia persica* subsp. *iranica* (Akhani) G. Kadereit & Piirainen, in Iran, and probably in eastern Mediterranean and Southwest Asia.

In North America:

- *Salicornia maritima* S.L.Wolff & Jefferies, Slender glasswort, coasts of southeastern Canada, northeastern United States, southern Alaska. (proved to be genetically identical to *Salicornia europaea*)
- *Salicornia bigelovii* Torrey, Dwarf glasswort, Gulf of Mexico, Atlantic coasts to Maine, southern California.
- *Salicornia depressa* Standley, (probable synonym *Salicornia virginica*). Pacific coasts from Alaska to California, Atlantic coasts from Canada to South Carolina.
- *Salicornia rubra* A.Nelson, salty inland habitats of the central United States and Canada.

In Africa:

- *Salicornia praecox* A.Chev., western Senegal
- *Salicornia senegalensis* A.Chev., western Senegal

- *Salicornia perrieri* A.Chev., East African coasts from Mozambique to KwaZulu-Natal, Madagascar, probably in Tanzania and Zanzibar
- *Salicornia pachystachya* Bunge ex Ungern-Sternb., East African coasts from southern Kenya to KwaZulu-Natal, Madagascar
- *Salicornia meyeriana* Moss, South African coasts from Vanrhynsdorp to Durban
- *Salicornia uniflora* Toelken, southwestern Namibia (Lüderitz) and northwestern South Africa (Darling)

In South Asia:

- *Salicornia brachiata* Roxb., Umari keerai, coasts of Ceylon and from East India to Bengal.

Culinary

Salicornia europaea is edible, either cooked or raw. In England, it is one of several plants known as **samphire** ; the term samphire is believed to be a corruption of the French name, *herbe de Saint-Pierre*, which means "St. Peter's herb".

Samphire is usually cooked, either steamed or microwaved, and then coated in butter or olive oil. Due to its high salt content, it must be cooked without any salt added, in plenty of water. It has a hard, stringy core, and after cooking, the edible flesh is pulled off from the core. This flesh, after cooking, resembles seaweed in color, and the flavor and texture are like young spinach stems or asparagus. Samphire is often used as a suitably maritime accompaniment to fish or seafood.

In addition to *S. europaea*, the seeds of *S. bigelovii* yield an edible oil. *S. bigelovii's* edibility is compromised somewhat because it contains saponins, which are toxic under certain conditions.

Umari keerai is cooked and eaten or pickled. It is also used as fodder for cattle, sheep and goats. In Kalpitiya, Sri Lanka, it is used to feed donkeys.

On the east coast of Canada, the plant is known as "samphire greens" and is a local delicacy. In Southeast Alaska, it is known as "beach asparagus". In Nova Scotia, Canada, they are known as "crow's foot greens". In British Columbia, Canada, they are known as "Sea Asparagus".In the United States, they are known as "sea beans" when used for culinary purposes. Other names include "sea asparagus", "sea green bean", "pousse-pierre", "passe-pierre", "pousse-pied", "sea pickle", and "marsh samphire".

In India, researchers at the Central Salt and Marine Chemicals Research Institute developed a process to yield culinary salt from *Salicornia brachiata.* The resulting product is known as vegetable salt and sold under the brand name Saloni.

Industrial Use

Historical

Salicornia virginica

The ashes of glasswort and saltwort plants and of kelp were long used as a source of soda ash (mainly sodium carbonate) for glassmaking and soapmaking. The introduction of the LeBlanc process for industrial production of soda ash superseded the use of plant sources in the first half of the 19th century.

Umari keerai is used as raw material in paper and board factories.

Salicornia europaea

Contemporary

Because *Salicornia bigelovii* can be grown using saltwater and its seeds contain high levels of unsaturated oil (30%, mostly linoleic acid) and protein (35%), it can be used to produce animal feedstuff and as a biofuel feedstock on coastal land where conventional crops cannot be grown. Adding nitrogen-based fertiliser to the seawater appears to increase the rate of growth and the eventual height of the plant, and the effluent from marine aquaculture (e.g. shrimp farming) is a suggested use for this purpose.

Experimental fields of *Salicornia* have been planted in Ras al-Zawr (Saudi Arabia), Eritrea (northeast Africa) and Sonora (northwest Mexico) aimed at the production of biodiesel. The company responsible for the Sonora trials (Global Seawater) claims between 225 and 250 gallons of BQ-9000 biodiesel can be produced per hectare (approximately 2.5 acres) of salicornia, and is promoting a $35 million scheme to create a 12,000-acre (49 km^2) salicornia farm in Bahia de Kino. Stems and roots of *Salicornia brachiata* plants had high cellulose contents (ca. 30%), whereas tender stem tips contents low cellulose content (9.2%). *Salicornia brachiata* revealed the dominance of rhamnose, arabinose, mannose, galactose, and glucose, with meager presence of ribose and xylose in their structural polysaccharide.

Environmental Uses

Pickleweed is used in Phytoextraction, it is highly effective at removing selenium from soil, which is absorbed by the plant and then released into the atmosphere to be dispersed by prevailing winds. Pickleweed (Salicornia bigelovii) has been found to have average volatilization rates 10-100 times higher than other species.

SPARTINA ALTERNIFLORA

Spartina alterniflora, the smooth cordgrass, saltmarsh cordgrass, or salt-water cord grass, is a perennial deciduous grass which is found in intertidal wetlands, especially estuarine salt marshes. It grows 1-1.5 m tall (3-5 feet,) and has smooth, hollow stems which bear leaves up to 20-60 cm (8 inches-2 feet) long and 1.5 cm (1/2 inch) wide at their base, which are sharply tapered and bend down at their tips. Like its relative saltmeadow cordgrass S. patens, it produces flowers and seeds on only one side of the stalk. The flowers are a yellowish-green, turning brown by the winter. It has rhizoidial roots, which, when broken off, can result in vegetative asexual growth. The roots are an important food resource for snow geese. It can grow in low marsh (frequently inundated by the tide) as well as high marsh (less frequently inundated), but it is usually restricted to low marsh because it is outcompeted by saltmeadow cordgrass in the high marsh. It grows in a wide range of salinities, from about 5 psu to marine (32 psu), and has been described as the "single most important marsh plant species in the estuary" of Chesapeake Bay. It is described as intolerant of shade.

S. alterniflora is noted for its capacity to act as an environmental engineer. It grows out into the water at the seaward edge of a salt marsh, and accumulates sediment and enables other habitat-engineering species, such as mussels, to settle. This accumulation of sediment and other substrate-building species gradually builds up the level of the land at the seaward edge, and other, higher-marsh species move onto the new land. As the marsh accretes, S. alterniflora moves still further out to form a new edge. S. alterniflora grows in tallest forms at the outermost edge of a given marsh, displaying shorter morphologies up onto the landward side of the Spartina belt.

S. alterniflora is native to the Atlantic coast of the Americas from Newfoundland, Canada, south to northern Argentina, where it forms a dominant part of brackish coastal saltmarshes.

Problems as an Invasive Species

Spartina alterniflora can become an invasive plant, either by itself or by hybridizing with native species and preventing propagation of the pure native strain. One example of an invasive Spartina alterniflora hybrid is that of S. anglica. S. anglica is a fertile polyploid derived from the hybrid S.alterniflora × townsendii (S. alterniflora × S. maritima), first found when American S. alterniflora was introduced to southern England in about 1870 and came into contact with the local native S. maritima. S. anglica has a variety of traits that allow it to outcompete native plants, including a high saline tolerance and the ability to perform photosynthesis at lower temperatures more productively than other similar plants. It can grow on a wider range of sediments than other species of Spartina, and can survive inundation in salt water for longer periods of time. S. anglica has since spread throughout northwest Europe, and (following introduction for erosion control) eastern North America.

In Willapa Bay of Washington state, Spartina alterniflora was probably an accidental introduction during oyster transplants during the nineteenth century. Presently, it covers 6,000-10,000 ha (15 to 25 thousand acres) of land, according to a 1999 estimate. In contrast, in the 1950s, under 160 ha (400 acres) of Spartina alterniflora were present. It is also making inroads into Puget Sound and Grays Harbor in Washington. The grass can hinder water circulation and drainage or block boating channels. Meadows of S. alterniflora can crowd out native species, reducing biodiversity and altering the environment; as a result of S. alterniflora's growth, invertebrates that live in mud flats disappear as their habitat is overgrown, and in turn, food sources shrink for birds who feed on those invertebrates.

In California, four species of exotic Spartina (S. alterniflora, S. densiflora, S. patens, and S. anglica) have been introduced to the San Francisco Bay region. Spartina alterniflora is well established in San Francisco Bay, and has had the greatest impact of all the cordgrasses in San Francisco Bay. It was introduced in 1973 by the Army Corps of Engineers in an attempt to reclaim marshland, and was spread and replanted around the bay in further restoration projects. It demonstrated an ability to outcompete the native S. foliosa, and to potentially eliminate it from San Francisco Bay.

Spartina alterniflora has also been found to hybridize with S. foliosa, producing offspring that may be an even greater threat than S. alterniflora by itself. The hybrid can physically modify the environment to the detriment of native species, and the hybrid populations have spread into creeks, bays, and more remote coastal locations. The hybrids produce enormous amounts of pollen, which swamp the stigmas of the native S. foliosa flowers to produce even larger numbers of hybrid offspring, leaving the native Spartina little chance to grow as a pure strain. The hybrids also produce much larger numbers of fertile seeds than the native Spartina, and are producing a hybrid population that is not only increasing in numbers, but increasing in its rate of population growth. The hybrids may also be able to fertilize themselves, which the native Spartina cannot do, and which increases the spread of the hybrid swarm even further.

Several means of control and eradication have been employed against Spartina alterniflora where it has become a pest. Hand pulling is effective, but only when done thoroughly and carefully, lest the plant's seeds spread to infest other areas. Glyphosate, an herbicide, is approved in Washington to kill it. In Willapa Bay, leafhopper bugs (Prokelisia marginata) have been employed to kill the plants, which threaten the oyster industry there. Surveys, by air, land, and sea are conducted in infested and threatened areas near San Francisco to determine Spartina's spread.

The caterpillars of Aaron's skipper (Poanes aaroni) have only been found on this species to date.

TETRAGONIA TETRAGONIOIDES

Tetragonia tetragonioides (previously *T. expansa*) is a leafy groundcover also known as **Botany Bay spinach**, **Cook's cabbage**, ***kōkihi*** (in Māori), **New Zealand spinach**, **sea spinach**, and **tetragon**. Its Australian names of **warrigal greens** and **warrigal cabbage** come from the local use of *warrigal* to describe plants that are wild (not farmed originally). It is native to Argentina, Australia, Chile, Japan, and New Zealand.

The species, rarely used by indigenous people as a leaf vegetable, was first mentioned by Captain Cook. It was immediately picked, cooked, and pickled to help fight scurvy, and taken with the crew of the *Endeavour*. It spread when the explorer and botanist Joseph Banks took seeds back to Kew Gardens during the latter half of the 18th century. For two centuries, *T. tetragonioides* was the only cultivated vegetable to have originated from Australia and New Zealand.

There are some indications that Māori did eat *k kihi* perhaps more regularly. "To counteract the bitterness of the older leaves of this herb, the Māori boiled it with the roots of the convolvulus (*p hue*)".

The species prefers a moist environment for growth. The plant has a trailing habit, and will form a thick carpet on the ground or climb though other vegetation and hang downwards. The leaves of the plant are 3–15 cm long, triangular in shape, and bright green. The leaves are thick, and covered with tiny papillae that look like waterdrops on the top and bottom of the leaves. The flowers of the plant are yellow, and the fruit is a small, hard pod covered with small horns. The plant is a halophyte and grows well in saline ground.

Cultivation

It is grown for the edible leaves, and can be used as food or an ornamental plant for ground cover. As some of its names signify, it has similar flavour and texture properties to spinach, and is cooked like spinach. Like spinach, it contains oxalates; its medium to low levels of oxalates need to be removed by blanching the leaves in hot water for one minute, then rinsing in cold water before cooking. It can be found as an invasive plant in North and South America, and has been cultivated along the East Asian rim. It thrives in hot weather, and is considered an heirloom vegetable. Few insects will bother it, and even slugs and snails do not seem to bother it.

The thick, irregularly-shaped seeds should be planted just after the last spring frost. Before planting, the seeds should be soaked for 12 hours in cold water, or 3 hours in warm water. Seeds should be planted 5–10 mm (0.2–0.4 in) deep, and spaced 15–30 cm (5.9–11.8 in) apart. The seedlings will emerge in 10–20 days, and it will continue to produce greens through the summer.

REFERENCES

1. Glenn, E. P.; et al. (1999). "Salt tolerance and crop potential of halophytes". Critical Review in Plant Sciences. 18 (2): 227–55. doi:10.1080/07352689991309207.
2. Glenn, E. P.; Brown, J. J.; O'Leary, J. W. (1998). "Irrigating Crops with Seawater", Scientific American, Vol. 279, no. 8, Aug. 1998, pp. 56-61.
3. Gupta, Bhaskar; Huang, Bingru (3 April 2014). "Mechanism of Salinity Tolerance in Plants: Physiological, Biochemical, and Molecular Characterization". International Journal of Genomics. doi:10.1155/2014/701596. Retrieved 15 October 2015.
4. "Fact Sheet: Alternative Fuels". IATA. December 2013. Retrieved 2014-01-28.
5. Glenn, Edward P.; Brown, J. Jed; O'Leary, James W. (August 1998). "Irrigating Crops with Seawater" (PDF). Scientific American. USA: Scientific American, Inc. (August 1998): 76–81 . Retrieved 2008-11-17.
6. The Plant List
7. Flowering Plants of the Santa Monica Mountains, Nancy Dale, 2nd Ed., 2000, p. 175
8. Medicinal Plants of the SW - Anemopsis californica, retrieved on July 17, 2007.
9. Boufford, D. E. (1997). "Anemopsis californica". In Flora of North America Editorial Committee. Flora of North America North of Mexico. 3. New York and Oxford: Oxford University Press. 9780195112467.
10. http://ucjeps.berkeley.edu/cgi-bin/get_JM_treatment.pl?7085,7086,7087 Jepson Manual Treatment
11. Quoted here by permission of the author: Soule, J. A. 2011. Father Kino's Herbs: Growing and Using Them Today. Tierra del Sol Press, Tucson, AZ.
12. G. Bentham. Plantas Hartwegianas.(1839)p. 30
13. Anemopsis californica - Plants For A Future database report, retrieved on July 17, 2007
14. Kay, M.A. Healing with Plants in the American and Mexican West. University of Arizona Press, Tucson, AZ, 1996.

15. Sencoval, Annette. Home Grown Healing Traditional Home Remedies from Mexico. Berkley Books, New York, NY, 1998.
16. Moore, Michael. Medicinal Plants of the Desert and Canyon West. Museum of New Mexico Press, Santa Fe, NM, 1989.
17. Soule, J. A. 2011. Father Kino's Herbs: Growing and Using Them Today. Tierra del Sol Press, Tucson, AZ.
18. Soule, J. A. 2011. Father Kino's Herbs: Growing and Using Them Today. Tierra del Sol Press, Tucson, AZ.
19. Bakker, Elna. "Yerba Mansa as Ground Cover." Pacific Horticulture 49 (4): 47-49. Pacific Horticultural Foundation, San Francisco, CA, 1988.
20. Sunset Western Garden Book, 1995:606–607
21. Quattrocchi, Umberto (2000). CRC World Dictionary of Plant Names. I: A-C. CRC Press. p. 232. ISBN 978-0-8493-2675-2.
22. Stanley L. Welsh: Atriplex - online, In: Flora of North America Editorial Committee (ed.): Flora of North America North of Mexico, Volume 4: Magnoliophyta: Caryophyllidae, part 1., Oxford University Press, New York. 2003, ISBN 0-19-517389-9.
23. Gelin Zhu, Sergei L. Mosyakin & Steven E. Clemants: Chenopodiaceae: Atriplex - online, In: Wu Zhengyi, Peter H. Raven, Deyuan Hong (ed.): Flora of China, Volume 5: Ulmaceae through Basellaceae., Science Press und Missouri Botanical Garden Press, Beijing und St. Louis, 2003, ISBN 1-930723-27-X.
24. Gudrun Kadereit, Evgeny V. Mavrodiev, Elizabeth H. Zacharias & Alexander P. Sukhorukov: Molecular phylogeny of Atripliceae (Chenopodioideae, Chenopodiaceae): Implications for systematics, biogeography, flower and fruit evolution, and the origin of C4 Photosynthesis. - American Journal of Botany 97(10): 1664-1687, 2010.
25. Christopher Evans (2015), North West Cambridge Archaeology: University of Cambridge 2013 Excavations, The Traveller's Rest Sub-site (PDF), Cambridge Archaeological Unit University Of Cambridge, pp. 100–113
26. Mistranslated as "mallows" in the King James Bible and as Nesseln (nettles) in the Luther Bible
27. PEARCE, KELLY & JACOB, ROBIN (2004): Saltbush lifts sheep meat vitamin content. Farming Ahead 153(October): 63. PDF fulltext Archived July 23, 2005, at the Wayback Machine.
28. NORMAN, HAYLEY C.; FREIND, COLBY; MASTERS, DAVID G.; RINTOUL, ALLAN J.; DYNES, ROBYN A. & WILLIAMS, IAN H. (2004): Variation within and between two saltbush species in plant composition and subsequent selection by sheep. Aust. J. Agr. Res. 55(9): 999–1007. doi:10.1071/AR04031 (HTML abstract)
29. Carl Linnaeus (1753), "Atriplex", Species Plantarum, 2, Lars Salvius/Biodiversity Heritage Library, pp. 1052–1054, retrieved 19 May 2015
30. Atriplex at Tropicos, accessed 2013-07-11

31. Tasmanian name, also transcribed trucanini, trucaninny, trugannini, trugernanna, etc. The plant was the namesake for Truganini, among the last of her people.
32. Elizabeth H. Zacharias, Bruce G. Baldwin (2010): A Molecular Phylogeny of North American Atripliceae (Chenopodiaceae), with Implications for Floral and Photosynthetic Pathway Evolution. In: Systematic Botany 35(4), p.839-857. doi:10.1600/036364410X539907
33. Dransfield, John; Natalie W. Uhl; Conny B. Asmussen; William J. Baker; Madeline M. Harley; Carl E. Lewis (2005). "A New Phylogenetic Classification of the Palm Family, Arecaceae". Kew Bulletin. 60 (4): 559–69. JSTOR 25070242.
34. Balick et al. 1987, Anderson and Balick 1988
35. Pintaud, Jean-Christophe (2008). "An overview of the taxonomy of Attalea (Arecaceae)" (PDF). Revista Peruana de Biología. 15 (supl. 1): 055–063.
36. Henderson, Andrew; Gloria Galeano; Rodrigo Bernal (1995). Field Guide to the Palms of the Americas. Princeton, New Jersey: Princeton University Press. pp. 153–164. ISBN 0-691-08537-4.
37. Anderson, Anthony B.; William L. Overal; Andrew Henderson (1988). "Pollination Ecology of a Forest-Dominant Palm (Orbignya phalerata Mart.) in Northern Brazil". Biotropica. 20 (3): 192–205. doi:10.2307/2388234. JSTOR 2388234.
38. First biofuel flight, BBC News, 24/2/08
39. May et al. (1985) Subsistence Benefits from the Babassu Palm (Orbignya martiana). Economic Botany, 39(2), p. 124 http://www.alice.cnptia.embrapa.br/bitstream/doc/955767/1/7295.pdf
40. Renold, Jaye (2015) Brazil's Warrior Women - Educational Pack: http://ifnotusthenwho.me/films/womens-movement-access-babassu-oil/
41. Silzer, Tanya (January 2000). "Panicum virgatum L., Switchgrass, prairie switchgrass, tall panic grass". Rangeland Ecosystems & Plants Fact Sheets. University of Saskatchewan Department of Plant Sciences. Retrieved 2007-12-08.
42. Secter, Bob. "Plentiful switch grass emerges as breakthrough biofuel". The San Diego Union-Tribune. Retrieved 2008-05-24.
43. Index of Species Information (2013). "Species: Panicum virgatum". US Forest Service. Archived from the original on 2012-11-08. Retrieved 2013-12-13.
44. Ernst Seed Catalog Web Page (2007). "Switchgrass and Warm Season Grass Planting Guide". Ernst Conservation Seeds. Archived from the original on 2007-08-03. Retrieved 2007-12-08.
45. "PLANTS Profile for Panicum virgatum (switchgrass)". United States Department of Agriculture, Natural Resources Conservation Service. Retrieved 2008-05-21.
46. Ball, D.M.; Hoveland, C.S.; Lacefield, G.D. (2002). Southern Forages (3rd ed.). International Plant Nutrition Institute. p. 26. ISBN 978-0-9629598-3-7.
47. Establishing and Managing Switchgrass as an Energy Crop, Forage and Grazinglands, 2008.

48. Switchgrass as a Bioenergy Crop, ATTRA - National Sustainable Agricultural Information Service, 2006.

49. "Aprovechamiento de recursos vegetales y animales para la produccion de biocombustibles" (PDF) (in Spanish). INTA. 26 June 2008.

50. David J. Parrish, John H. Fike, David I. Bransby, Roger Samson. Establishing and Managing Switchgrass as an Energy Crop. Forage and Grazinglands. 2008

51. "Archived copy" (PDF). Archived from the original (PDF) on 2009-01-22. Retrieved 2009-03-16.

52. Samson, R. (2007). "Switchgrass Production in Ontario: A Management Guide" (PDF). Resource Efficient Agricultural Production (REAP) - Canada. Retrieved 2008-05-24.

53. Bransby, David (2005). "Switchgrass Profile". Bioenergy Feedstock Information Network (BFIN), Oak Ridge National Laboratory. Retrieved 2008-05-24.

54. Samson, R., Bailey-Stamler, S., & Ho Lem, C. Optimization of Switchgrass Management for Commercial Fuel Pellet Production (Final report prepared by REAP-Canada for the Ontario Ministry of Food, Agriculture and Rural Affairs (OMAFRA) under the Alternative Renewable Fuels Fund). 2008

55. McLaughlin, S.B., Kzos, L.A. Development of switchgrass (Panicum virgatum) as a bioenergy feedstock in the United States. Biomass and Bioenergy. Volume 28. 2005. 515-535.

56. Dale, B.; Kim, S. (2004). "Cumulative Energy and Global Warming Impact from the Production of Biomass for Biobased Products" (PDF). Journal of Industrial Ecology. 7 (3-4): 147–162.

57. Samson, R.; et al. (2005). "The Potential of C4 Perennial Grasses for Developing a Global BIOHEAT Industry" (PDF). Critical Reviews in Plant Science. 25: 461–495.

58. Samson, R. et al. Developing Energy Crops for Thermal Applications: Optimizing Fuel Quality, Energy Security and GHG Mitigation. In Biofuels, Solar and Wind as Renewable Energy Systems: Benefits and Risks. D. Pimental. (Ed.) Springer Science, Berlin, Germany. 2008. 395-423.

59. Adrienne Mand Lewin (2006-02-01). "Switchgrass: The Super Plant Savior?". ABC News.

60. "Switch Grass: Alternative Energy Source?". National Public Radio. 2006-02-01.

61. Dana Bash; Suzanne Malveaux; et al. (2006-02-01). "Bush has plan to end oil 'addiction'". CNN.

62. "Since that mention in the 2006, investment in switch grass has exploded, thanks in large part, experts say, to the President's speech. Venture capitalists have poured over $100 million into private companies that are exploring the technology necessary to convert switchgrass into fuel, and large publicly owned companies are also directing their research dollars into biofuels."Jessica Yellin; Katie Hinman; Nitya Venkataraman (2007-01-23). "What Happened to Bush Call for Switchgrass?". ABC News.

63. "Switchgrass: Native American Powerhouse?". Renewable Energy Resources. Retrieved 2007-01-05.
64. Iogen Corporation Iogen Corporation, 2009.
65. Farrell, A.E. Ethanol Can Contribute to Energy and Environmental Goals. Science. Volume 311. 2006. 506-508.
66. Ethanol Across America;The Impact of Ethanol Production on Food, Feed and Fuel
67. Ethanol Producer Magazine; Tiny Bubbles to Make You Happy
68. Ethanol Producer Magazine; Study explores increased ethanol yield
69. Ethanol Producer Magazine; Genencor enzyme reduces dosing 3-fold
70. Oleg Kozyuk;Arisdyne Systems Inc.; US patent US 7,667,082 B2; Apparatus and Method for Increasing Alcohol Yield from Grain
71. "Switch Grass: Alternative Energy Source?". NPR. Retrieved 2007-01-05.
72. Wang, M., Wu, M., & Huo, H. Life-cycle energy and greenhouse gas emission impacts of different corn ethanol plant types. Environmental Research Letters. Volume 2. 2007. 1-13.
73. M. R. Schmer; K. P. Vogel; R. B. Mitchell; R. K. Perrin (2008). "Net energy of cellulosic ethanol from switchgrass". PNAS. 105: 464–469. doi:10.1073/pnas.0704767105. PMID 18180449.
74. Lynd, L. R.; Cushman, J. H.; Nichols, R. J.; Wyman, C. E. (1991). "Fuel Ethanol from Cellulosic Biomass". Science. 251 (4999): 1318–1323. doi:10.1126/science.251.4999.1318.
75. http://www.iowaswitchgrass.com/
76. Show Me Energy Cooperative
77. Bailey Stamler, S., R. Samson and C. Ho Lem. Biomass resources options: Creating a BIOHEAT supply for the Canadian greenhouse industry. Final report to Natural Resources Canada, Ottawa. 2006. 38 pages
78. Samson, R., and S. Bailey Stamler. Going Green for Less: Cost-Effective Alternative Energy Sources. C.D. Howe Institute Commentary 282. Economic Growth and Innovation. 2009. 25 pages
79. "Samson et al. 2007": http://www.reap-canada.com/online_library/feedstock_biomass/The%20Emerging%20Agro-Pellet%20Industry%20in%20Canada%20(Samson%20et%20al.,%202007).pdf R. Samson, S. Bailey and C. Ho Lem. The Emerging Agro-Pellet Industry in Canada. 2007
80. "Samson et al., 2008": In regions where the potassium and chlorine contents of switchgrass cannot be successfully leached out for thermal applications, it may be that biogas applications for switchgrass will prove more promising. Switchgrass has demonstrated some promise in biogas research as an alternative feedstock to whole plant corn silage for biogas digesters.
81. "Frigon et al. 2008 http://www.gtmconference.ca/site/downloads/2008presentations/5B3%20-Frigon.pdf J.C. Frigon, P. Mehta, S.R.

Guiot. The bioenergy potential from the anaerobic digestion of switchgrass and other energy crops. National Research Council Canada. Growing the Margins Conference: Energy, Bioproducts and Byproducts from farms and food sectors. April 2–5, 2008, London, Ontario

82. "plentiful_switchgrass"
83. Bai, Y., Luo, L, & van der Voe, E. (2010). Life cycle assessment of switchgrass-derived ethanol as transport fuel. Int J Life Cycle Assess. 15 (15), 468-477. This can be found on-line at: http://www.springerlink.com/content/ap35w2q46420tn4x/fulltext.pdf.
84. "Bacteria take plants to biofuel in one step". Science News. Retrieved 2014-06-04.
85. "Direct conversion of plant biomass to ethanol by engineered Caldicellulosiruptor bescii". Pnas.org. Retrieved 2014-06-04.
86. Biomass Combo, Chemical & Engineering News, 86, 33, 18 Aug. 2008, p. 13
87. Plant Biotechnology Journal, 2008, 6, 663
88. "Plant Fact Sheet, Panicum virgatum (switchgrass)" (PDF). United States Department of Agriculture, Natural Resources Conservation Service, Plant Materials Program. 2006-05-06. Retrieved 2008-05-21.
89. "KDOT Bid Tabs". Kansas Department of Transportation. Retrieved 2008-05-20.
90. Lee, S.T.; Stegelmeier, B.L.; Gardner, D.R.; Vogel, K.P. (2001). "The isolation and identification of steroidal sapogenins in switchgrass". J Nat Toxins. 10 (4): 273–81. PMID 11695816.
91. Johnson, A.L.; Divers, T.J.; Freckleton, M.L.; McKenzie, H.C.; Mitchell, E.; Cullen, J.M.; McDonough, S.P. (2006). "Fall Panicum (Panicum dichotomiflorum) Hepatotoxicosis in Horses and Sheep". Journal of Veterinary Internal Medicine. 20 (6): 1414–1421. doi:10.1892/0891-6640(2006)20[1414:FPPDHI]2.0.CO;2. PMID 17186859. Retrieved 2008-05-24.
92. Stegelmeier, B.L.; Elmore, S.A.; Lee, S.T.; James, L.F.; Gardner, D.R.; Panter, K.E.; Ralphs, M.H.; Pfister, J.A. (2007). "Switchgrass (panicum Virgatum) Toxicity In Rodents, Sheep, Goats And Horses". Poisonous Plant Global Research and Solutions. 19: 113–117. Retrieved 2008-05-24.
93. Ball, D.M.; Hoveland, C.S.; Lacefield, G.D. (2006). "Table 28. Guidelines for rotational stocking of selected forage crops". Forage Crop Pocket Guide. International Plant Nutrition Institute.
94. Ball, D.M.; Hoveland, C.S.; Lacefield, G.D. (2006). "Table 33b. Total Digestible Nutrients (TDN) and Relative Feed Value (RFV) Ranges for Various Forage Crops". Forage Crop Pocket Guide. International Plant Nutrition Institute.
95. Wolf, D.D.; Fiske, D.A. (1995). "Planting and managing switchgrass for forage, wildlife, and conservation" (PDF). Virginia Cooperative Extension Publication: 418–013. Retrieved 2008-05-24.
96. Leimbach, D.D. &Marcello, G.J. (2015). Observed effects of warm season perennial grass agriculture on resident mammal species. Sustainable Agriculture Research,4(2):70-77.

97. Hipple, Patricia C.; Duffy, Michael D. (2002). "Farmers' Motivations for Adoption of Switchgrass" (PDF). In Jules Janick, Anna Whipkey. Trends in New Crops and New Uses. Fifth National Symposium, New Crops and New Uses, Strength in Diversity. Alexandria, Virginia: American Society for Horticultural Science. pp. 252–266. ISBN 978-0-09-707565-5. Retrieved 2008-05-23.
98. Bies, Laura (2006-11-01). "The Biofuels Explosion: Is Green Energy Good for Wildlife?". Wildlife Society Bulletin. 34 (4): 1203–1205. doi:10.2193/0091-7648(2006)34[1203:TBEIGE]2.0.CO;2.
99. Salicornia, Integrated Taxonomic Information System, serial number 20646.
100. Ball, Peter W. (2004). "Salicornia L.," in Flora of North America: North of Mexico Volume 4: Magnoliophyta: Caryophyllidae, part 1, Editorial Committee of the Flora of North America (Oxford University Press, 2004). ISBN 978-0-19-517389-5. Online version retrieved August 10, 2016.
101. Kadereit, G.; Borsch, T.; Weising, K.; Freitag, H. (2003). "Phylogeny of Amaranthaceae and Chenopodiaceae and the evolution of C4 photosynthesis". International Journal of Plant Sciences. 164 (6): 959–986. doi:10.1086/378649.
102. Kadereit, G.; Ball, P.; Beer, S.; Mucina, L.; Sokoloff, D.; Teege, P.; Yaprak, A.E.; Freitag, H. (2007). "A taxonomic nightmare comes true: phylogeny and biogeography of glassworts (Salicornia L., Chenopodiaceae)". Taxon. 56 (4): 1143–1170. doi:10.2307/25065909.
103. Carl Linnaeus (1753). Species Plantarum, Tomus I: 3. First description of Salicornia, scanned at BHL
104. "Salicornia". Tropicos. Missouri Botanical Garden. Retrieved 2016-08-11.
105. Kadereit, G.; Piirainen, M.; Lambinon, J.; Vanderpoorten, A. (2012). "Cryptic taxa should have names. Reflections on the glasswort genus Salicornia (Amaranthaceae)". Taxon. 61: 1227–1239.
106. "Salicornia", page of the Plants for a Future website. Retrieved July 14, 2007.
107. Davidson, Alan (2002). The Penguin Companion To Food (Penguin), p. 828. ISBN 978-0-14-200163-9. On Food and Cooking: The Science and Lore of the Kitchen, Completely Revised and Updated (Scribner, New York), p. 317. ISBN 978-0-684-80001-1.
108. Salicornia, oil-yielding plant for coastal belts, The Hindu
109. http://westcoastseaweed.com/
110. Cook's Thesaurus: Sea Vegetables, retrieved 2012-10-08.
111. "Indian scientists produce salt from vegetable". The Economic Times (India). 18 May 2003.
112. Glenn, Edward P.; Brown, J. Jed; O'Leary, James W. (August 1998). "Irrigating Crops with Seawater" (PDF). Scientific American. USA: Scientific American, Inc. (August 1998): 76–81. Retrieved 2008-11-17.
113. Clark, Arthur (November–December 1994). "Samphire: From Sea to Shining Seed" (PDF). Saudi Aramco World. Saudi Aramco. Retrieved 2008-11-17.

114. Alsaeedi, Abdullah H. (2003 (1424H)). "Di Pattern of Salicornia Vegetative Growth in Relation to Fertilization" (PDF). Journal of King Faisal University. Al-Hassa: King Faisal University. 4 (1): 105–118. Retrieved 2008-11-17. adequate fertilization increases significantly the relative growth rate especially during the 'rapid' phase of the vegetative stage Check date values in: |date= (help)

115. "USIJI Uniform Reporting Document" (PDF). United States Initiative on Joint Implementation (USIJI). c. 1997. Retrieved 2008-11-17. Project Salicornia: Halophyte Cultivation in Sonora External link in |publisher= (help)

116. Ryan C. Christiansen (2008-07-31). "Sea asparagus can be oilseed feedstock for biodiesel". Biomass Magazine. Retrieved 2008-11-17.

117. Dickerson, Marla (2008-07-10). "Letting the sea cultivate the land". Los Angeles Times. Tribune Company. Retrieved 17 November 2008.

118. Sanandiya, Naresh D.; Prasad, Kamalesh; Meena, Ramavatar; Siddhanta, Arup K. (2010-04-01). "Cellulose of Salicornia brachiata". Natural Product Communications. 5 (4): 603–606. ISSN 1934-578X. PMID 20433080.

119. Sanandiya, Naresh D.; Siddhanta, A. K. (2014-11-04). "Chemical studies on the polysaccharides of Salicornia brachiata". Carbohydrate Polymers. 112: 300–307. doi:10.1016/j.carbpol.2014.05.072. ISSN 1879-1344. PMID 25129748.

120. Eichhorn, Peter H. Raven ; Ray F. Evert ; Susan E. (2011). Biology of plants (8. ed.). New York, NY: Freeman. p. 12. ISBN 9781429219617.

121. Terry, N; Zayed, AM; De Souza, MP; Tarun, AS (June 2000). "SELENIUM IN HIGHER PLANTS." (PDF). Annual Review of Plant Physiology and Plant Molecular Biology. 51: 401–432. doi:10.1146/annurev.arplant.51.1.401. PMID 15012198.

122. "Spartina alterniflora". Natural Resources Conservation Service PLANTS Database. USDA. Retrieved 24 November 2015.

123. http://www.fs.fed.us/database/feis/plants/graminoid/spaalt/all.html USDA Forest Service Fire Effects Information System (FEIS) for Spartina alterniflora

124. Lippson, AJ & RL Lippson. 2006. Life in the Chesapeake Bay, 3rd ed., p.295. Johns Hopkins Press.

125. http://plants.usda.gov/java/profile?symbol=spal USDA NRCS PLANTS profile for Spartina alterniflora

126. Ayres, D. R., D. L. Smith, K. Zaremba, S. Klohr, and D. R. Strong. 2004. Spread of exotic cordgrasses and hybrids (Spartina sp.) in the tidal marshes of San Francisco Bay, California, USA. Biological Invasions 6: 221-231. Accessed 7/24/12 at http://lamar.colostate.edu/~csbrown/BI528_2007/Genetics/Ayres%20et%20al%202004BioInv.pdf

127. Callaway, J.C., and M. N. Josselyn. 1992. The introduction and spread of smooth cordgrass (Spartina alterniflora) in South San Francisco Bay. Estuaries 15: 218-226.

128. Anttila, C. K., R. A. King, C. Ferris, D. R. Ayres, and D. R. Strong. 2000. Reciprocal hybrid formation of Spartina in San Francisco Bay. Molecular Ecology

9: 765-770. Accessed 7/24/12 at http://onlinelibrary.wiley.com/doi/10.1046/j.1365-294x.2000.00935.x/pdf

129. Ayres D.R.; Garcia-Rossi D.; Davis H.G.; Strong D.R. 1999. Extent and degree of hybridization between exotic (Spartina alterniflora) and native (S. foliosa) cordgrass (Poaceae) in California, USA determined by random amplified polymorphic DNA (RAPDs). Molecular Ecology 8: 1179-1186. Accessed 7/24/12 at http://onlinelibrary.wiley.com/doi/10.1046/j.1365-294x.1999.00679.x/pdf
130. "BSBI List 2007". Botanical Society of Britain and Ireland. Archived from the original (xls) on 2015-02-25. Retrieved 2014-10-17.
131. "Tetragonia tetragonioides". The Royal Botanic Gardens and Domain Trust. Retrieved 10 February 2014.
132. The New Oxford American Dictionary (3rd ed.). 2012. adjective (of a plant) not cultivated: warrigal melons.
133. Low, T., Wild Food Plants of Australia, Angus & Robertson, 1991, ISBN 0-207-16930-6
134. Riley, Murdoch (1994). "Maori Healing and Herbal". Paraparaumu, New Zealand: Viking Sevenseas N.Z. Ltd. pp. 7–10. ISBN 0854670955.
135. "Māori Healing and Herbal - New Zealand Ethnobotanical Sourcebook". Viking Sevenseas NZ Ltd. 1994. p. 221.
136. "Hungry? Try some bush tucker" (movie). The Sydney Morning Herald. 2011-06-28. Retrieved 2011-06-28.

CHAPTER 7

Frost and Heavy Metal Tolerance in Plants

Chapter Contents

Frost is water vapor, or water in gas form, that becomes solid. Frost usually forms on objects like cars, windows, and plants that are outside in air that is saturated, or filled, with moisture. Areas that have a lot of fog often have heavy frosts.

INTRODUCTION

Frost is the coating or deposit of ice that may form in humid air in cold conditions, usually overnight. In temperate climates it most commonly appears as fragile white crystals or frozen dew drops near the ground, but in cold climates it occurs in a greater variety of forms. Frost is composed of delicate branched patterns of ice crystals formed as the result of fractal process development.

Frost is known to damage crops or reduce future crop yields, therefore farmers in those regions where frost is a problem often invest substantial means to prevent its formation.

Frost forms when the temperature of a solid surface in an open area drops below the freezing point of water, forming clear crystalines In most temperate countries such temperatures usually are the result of heat loss by radiation at night, so those types of frost sometimes are called radiation frost.

Types of frost include crystalline hoar frost from deposition of water vapor from air of low humidity, white frost in humid conditions, window frost on glass surfaces, advection frost from cold wind over cold surfaces, black frost without visible ice at low temperatures and very low humidity, and rime under supercooled wet conditions.

The size of frost crystals varies depending on the time they have been building up and the amount of water vapor available. Frost crystals may be clear or translucent, but, like snow, a mass of frost crystals will scatter light in all directions, so that a coating of frost appears white.

Formation

Frost in the highest town in Venezuela, Apartaderos. Because of its location in an alpine tundra ecosystem called páramo, there exists a daily freeze-and-thaw cycle, sometimes described as "summer every day and winter every night."

If a solid surface is chilled below the dew point of the surrounding humid air and the surface itself is colder than freezing, ice will form on it. If the water deposits as a liquid that then freezes, it forms a coating that may look glassy, opaque, or crystalline, depending on its type. Depending on context, that process also may be called atmospheric icing. The ice it produces differs in some ways from crystalline frost, which consists of spicules of ice that typically project from the solid surface on which they grow.

The main difference between the ice coatings and frost spicules arises from the fact that the crystalline spicules grow directly from desublimation of water vapour from air, and desublimation is not a factor in icing of freezing surfaces. For desublimation to proceed the surface must be below the frost point of the air, meaning that it is sufficiently cold for ice to form without passing through the liquid phase. The air must be humid, but not sufficiently humid to permit the condensation of liquid water, or icing will result instead of desublimation. The size of the crystals depends largely on the temperature, the amount of water vapor available, and how long they have been growing undisturbed.

As a rule, except in conditions where supercooled droplets are present in the air, frost will form only if the deposition surface is colder than the surrounding air. For instance frost may be observed around cracks in cold wooden sidewalks when humid air escapes from the warmer ground beneath. Other objects on which frost commonly forms are those with low specific heat or high thermal emissivity, such as blackened metals; hence the accumulation of frost on the heads of rusty nails.

The apparently erratic occurrence of frost in adjacent localities is due partly to differences of elevation, the lower areas becoming colder on calm nights. Where static air settles above an area of ground in the absence of wind, the absorptivity and specific heat of the ground strongly influence the temperature that the trapped air attains.

Types

Hoar Frost

A spider web covered in air hoar frost

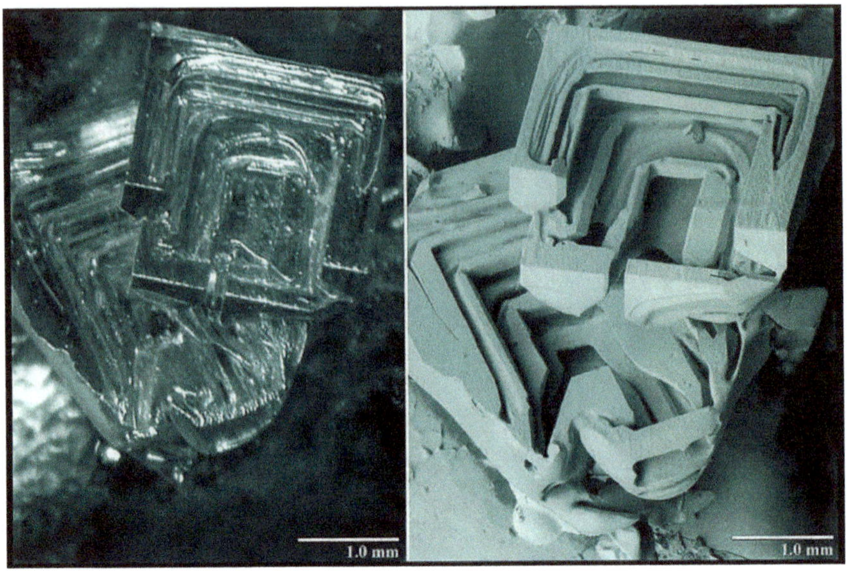

Depth hoar, imaged with light and with scanning electron microscopy.

Hoar frost (also hoarfrost, radiation frost, or pruina) refers to white ice crystals, deposited on the ground or loosely attached to exposed objects such as wires or leaves. They form on cold, clear nights when conditions are such that heat radiates out to the open sky faster than it can be replaced from nearby sources such as wind or warm objects. Under suitable circumstances, objects cool to below the frost point of the surrounding air, well below the freezing point of water. Such freezing may be promoted by effects such as flood frost or frost pocket. These occur when ground-level radiation losses cool air till it flows downhill and accumulates in pockets of very cold air in valleys and hollows. Hoar frost may freeze in such low-lying cold air even when the air temperature a few feet above ground is well above freezing.

The name **hoar** comes from an Old English adjective that means "showing signs of old age"; in this context it refers to the frost that makes trees and bushes look like white hair.

Hoar frost may have different names depending on where it forms:

- **air hoar** is a deposit of hoar frost on objects above the surface, such as tree branches, plant stems, and wires;
- **surface hoar** refers to fern-like ice crystals directly deposited on snow, ice or already frozen surfaces;
- **crevasse hoar** consists of crystals that form in glacial crevasses where water vapour can accumulate under calm weather conditions;
- **depth hoar** refers to faceted crystals that have slowly grown large within cavities beneath the surface of banks of dry snow. Depth hoar crystals grow continuously at the expense of neighbouring smaller crystals, so typically are visibly stepped and have faceted hollows.

When surface hoar covers sloping snowbanks, the layer of frost crystals may create an avalanche risk; when heavy layers of new snow cover the frosty surface, furry crystals standing out from the old snow hold off the falling flakes, forming a layer of voids that prevent the new snow layers from bonding strongly to the old snow beneath. Ideal conditions for hoarfrost to form on snow are cold clear nights, with very light, cold air currents conveying humidity at the right rate for growth of frost crystals. Wind that is too strong or warm destroys the furry crystals, and thereby may permit a stronger bond between the old and new snow layers. However, if the winds are strong enough and cold enough to lay the crystals flat and dry, carpeting the snow with cold, loose crystals without removing or destroying them or letting them warm up and become sticky, the frost interface between the snow layers still may present an avalanche danger, because the texture of the frost crystals differs from the snow texture and the dry crystals will not stick to fresh snow. Such conditions still prevent a strong bond between the snow layers.

In very low temperatures where fluffy surface hoar crystals form without subsequently being covered with snow, strong winds may break them off, forming a dust of ice particles and blowing them over the surface. The ice dust then may form yukimarimo, as has been observed in parts of Antarctica, in a process similar to the formation of dust bunnies and similar structures.

Hoar frost and white frost also occurs in man-made environments such as in freezers or industrial cold storage facilities. If such cold spaces or the pipes serving them are not well insulated and are exposed to ambient humidity, the moisture will freeze instantly depending on the freezer temperature. The frost may coat pipes thickly, partly insulating them, but such inefficient insulation still is a source of heat loss.

Advection Frost

A flower with advection frost on the tips of its petals.

Advection frost (also called **wind frost**) refers to tiny ice spikes that form when there is a very cold wind blowing over branches of trees, poles and other surfaces. It looks like rimming on the edge of flowers and leaves and usually it forms against the direction of the wind. It can occur at any hour, day or night.

Window Frost

Window frost (also called fern frost or ice flowers) forms when a glass pane is exposed to very cold air on the outside and warmer, moderately moist air on the inside. If the pane is not a good insulator (for example, if it is a single pane window), water vapour condenses on the glass forming frost patterns. With very low temperatures outside, frost can appear on the bottom of the window even with double pane energy efficient windows because the air convection between two panes of glass ensures that the bottom part of the glazing unit is colder than the top part. On unheated motor vehicles the frost will usually form on the outside surface of the glass first. The glass surface influences the shape of crystals, so imperfections, scratches, or dust can modify the way ice nucleates. The patterns in window frost form a fractal with a fractal dimension greater than one but less than two. This is a consequence of the nucleation process being constrained to unfold in two dimensions, unlike a snowflake which is shaped by a similar process but forms in three dimensions and has a fractal dimension greater than two.

If the indoor air is very humid, rather than moderately so, water will first condense in small droplets and then freeze into clear ice.

Similar patterns of freezing may occur on other smooth vertical surfaces, but they seldom are as obvious or spectacular as on clear glass.

White Frost

White frost is a solid deposition of ice which forms directly from water vapour contained in air.

White frost forms when there is a relative humidity above 90% and a temperature below −8 °C (18 °F) and it grows against the wind direction, since air arriving from windward has a higher humidity than leeward air, but the wind must not be strong or it damages the delicate icy structures as they begin to form. White frost resembles a heavy coating of hoar frost with big, interlocking crystals, usually needle-shaped.

Rime

Rime is a type of ice deposition that occurs quickly, often under conditions of heavily saturated air and windy conditions. Technically speaking, it is not a type of frost, since usually supercooled water drops are involved, in contrast to the formation of hoar frost, in which water vapour desublimates slowly and directly. Ships travelling through Arctic seas may accumulate large quantities of rime on the rigging. Unlike hoar frost, which has a feathery appearance, rime generally has an icy solid appearance.

Black Frost

Black frost (or "killing frost") is not strictly speaking frost at all, because it is the condition seen in crops when the humidity is too low for frost to form, but the temperature falls so low that plant tissues freeze and die, becoming blackened, hence the term "black frost". Black frost often is called "killing frost" because white frost tends to be less cold, partly because the latent heat of freezing of the water reduces the temperature drop.

Effect on Plants

Overview

Frost on the grass of a public park in January.

Many plants can be damaged or killed by freezing temperatures or frost. This varies with the type of plant, the tissue exposed, and how low temperatures get: a "light frost" of −2 to 0 °C (28 to 32 °F) will damage fewer types of plants than a "hard frost" below −2 °C (28 °F).

Plants likely to be damaged even by a light frost include vines—such as beans, grapes, squashes, melons—along with nightshades such as tomatoes, eggplants and peppers. Plants that may tolerate (or even benefit) from frosts include:

- root vegetables (e.g. beets, carrots, parsnips, onions)
- leafy greens (e.g. lettuces, spinach, chard, cucumber)
- cruciferous vegetables (e.g. cabbages, cauliflower, bok choy, broccoli, Brussels sprouts, radishes, kale, collard, mustard, turnips, rutabagas)

Even those plants that tolerate frost may be damaged once temperatures drop even lower (below −4 °C or 25 °F). Hardy perennials, such as *Hosta*, become dormant after the first frosts and regrow when spring arrives. The entire visible plant may turn completely brown until the spring warmth, or may drop all of its leaves and flowers, leaving the stem and stalk only. Evergreen plants, such as pine trees, withstand frost although all or most growth stops. Frost crack is a bark defect caused by a combination of low temperatures and heat from the winter sun.

Vegetation is not necessarily damaged when leaf temperatures drop below the freezing point of their cell contents. In the absence of a site nucleating the formation of ice crystals, the leaves remain in a supercooled liquid state, safely reaching temperatures of −4 to −12 °C (25 to 10 °F). However, once frost forms, the leaf cells may be damaged by sharp ice crystals. Hardening is the process by which a plant becomes tolerant to low temperatures. See also cryobiology.

Certain bacteria, notably *Pseudomonas syringae*, are particularly effective at triggering frost formation, raising the nucleation temperature to about −2 °C (28 °F).Bacteria lacking ice nucleation-active proteins (ice-minus bacteria) result in greatly reduced frost damage.

Protection Methods

Curitiba (Southern Brazil) is the coldest of Brazil's state capitals; the greenhouse of the Botanical Garden of Curitiba protects sensitive plants.

Typical measures to prevent frost or reduce its severity include one or more of:

- deploying powerful blowers to simulate wind, thereby preventing the formation of accumulations of cold air. There are variations on this theme. One of them is the selective inverted sink prevents frost by drawing cold air from the ground and blowing it up through a chimney. It was originally developed to prevent frost damage to citrus fruits in Uruguay. In New Zealand, helicopters are used in a similar function, especially in the vineyard regions like Marlborough. By dragging down warmer air from the inversion layers, and preventing the ponding of colder air on the ground, the low-flying helicopters prevent damage to the fruit buds. As the operations are conducted at night, and have in the past involved up to 130 aircraft per night in one region, safety rules are strict. Although not a dedicated method, wind turbines have similar (small) effect of vertically mixing air layers of different temperature.
- for high-value crops, farmers may wrap trees and cover crops.
- heating to slow the drop in temperature. This is not practical except for high value crops grown over small areas.
- production of smoke to reduce cooling by radiation
- spraying crops with a layer of water that releases latent heat, preventing harmful freezing of the tissues of the plants that it coats.

Such measures need to be applied with discretion, because they may do more harm than good; for example, spraying crops with water can cause damage if the plants become overburdened with ice. An effective low cost method for small crop farms and plant nurseries, exploits the latent heat of freezing. A pulsed irrigation timer delivers water through existing overhead sprinklers at a low volumes to combat frosts down to −5 °C (23 °F). If the water freezes it giving off its latent heat, preventing the temperature of the foliage from falling much below zero.

Frost-Free Areas

Frost-free areas are found mainly in the tropics, where they cover almost all land except at altitudes above about 3,000 metres or 9,800 feet near the equator and around 2,000 metres or 6,600 feet in the semi-arid middle tropics, but also in areas with subtropical climates that have winters tempered by strong oceanic influences. The most poleward frost-free areas are the lower altitudes of the Azores, Île Amsterdam, Île Saint-Paul, and Tristan da Cunha.

The only reliably frost-free areas in the contiguous United States are the Florida Keys and the coastal areas of the Channel Islands of California.

Personifications

Frost is personified in Russian culture as Ded Moroz. Indigenous peoples of Russia such as the Mordvins have their own traditions of frost deities.

English folklore tradition holds that Jack Frost, an elfish creature, is responsible for feathery patterns of frost found on windows on cold mornings.

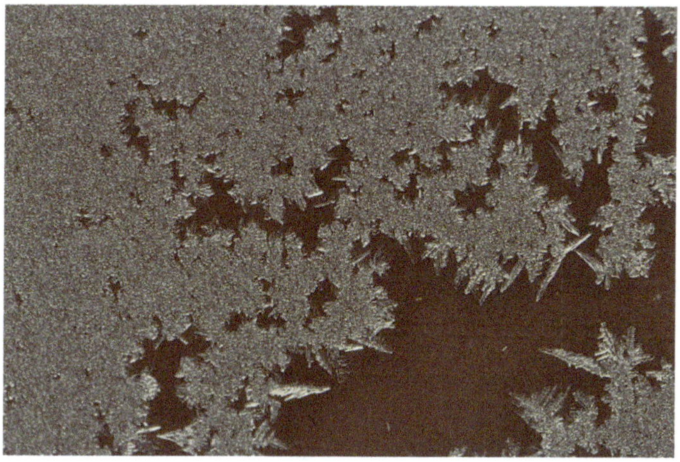

PLANT STRESS MEASUREMENT

Plant stress measurement is the quantification of environmental effects on plant health. When plants are subjected to less than ideal growing conditions, they are considered to be under stress. Stress factors can affect growth, survival and crop yields. Plant stress research looks at the response of plants to limitations and excesses of the main abiotic factors (light, temperature, water and nutrients), and of other stress factors that are important in particular situations (e.g. pests, pathogens, or pollutants). Plant stress measurement usually focuses on taking measurements from living plants. It can involve visual assessments of plant vitality, however, more recently the focus has moved to the use of instruments and protocols that reveal the response of particular processes within the plant (especially, photosynthesis, plant cell signalling and plant secondary metabolism)

- Determining the optimal conditions for plant growth, e.g. optimising water use in an agricultural system
- Determining the climatic range of different species or subspecies
- Determining which species or subspecies are resistant to a particular stress factor

Instruments Used to Measure Plant Stress

Measurements can be made from living plants using specialised equipment. Among the most commonly used instruments are those that measure parameters related to photosynthesis (chlorophyll content, chlorophyll fluorescence, gas exchange) or water use (porometer, pressure bomb). In addition to these general purpose instruments, researchers often design or adapt other instruments tailored to the specific stress response they are studying.

Photosynthesis Systems

Photosynthesis systems use infrared gas analyzers (IRGAS) for measuring photosynthesis. CO_2 concentration changes in leaf chambers are measured to provide carbon assimilation values for leaves or whole plants. Research has shown that the rate of photosynthesis is directly related to the amount of carbon assimilated by the plant. Measuring CO_2 in the air, before it enters the leaf chamber, and comparing it to air measured for CO_2 after it leaves the leaf chamber, provides this value using proven equations. These systems also use IRGAs, or solid state humidity sensors, for measuring H_2O changes in leaf chambers. This is done to measure leaf transpiration, and to correct CO_2 measurements. The light absorption spectrum for CO_2 and H_2O overlap somewhat, therefore, a correction is necessary for reliable CO_2 measuring results. The critical measurement for most plant stress measurements is designated by "A" or carbon assimilation rate. When a plant is under stress, less carbon is assimilated.CO_2 IRGAs are capable of measuring to approximately +/- 1 μmol or 1ppm of CO_2.

Because these systems are effective in measuring carbon assimilation and transpiration at low rates, as found in stressed plants, they are often used as the standard to compare to other types of instruments. Photosynthesis instruments come in field portable and laboratory versions. They are also designed to measure ambient environmental conditions, and some systems offer variable microclimate control of the measuring chamber. Microclimate control systems allow adjustment of the measuring chamber temperature, CO_2 level, light level, and humidity level for more detailed investigation.

The combination of these systems with fluorometers, can be especially effective for some types of stress, and can be diagnostic, e.g. in the study of cold stress and drought stress.

Chlorophyll Fluorometers

Chlorophyll fluorescence emitted from plant leaves, gives an insight into the health of the photosynthetic systems within the leaf. Chlorophyll fluorometers are designed to measure variable fluorescence of photosystem II. This variable fluorescence can be used to measure the level of plant stress. The most commonly used protocols include those aimed at measuring the photosynthetic efficiency of photosystem II, both in the light (ΔF/Fm') and in a dark-adapted state (Fv/Fm). Chlorophyll fluorometers are, for the most part, less expensive tools than photosynthesis systems, they also have a faster measurement time and tend to be more portable. For these reasons they have become one of the most important tools for field measurements of plant stress.

Fv/Fm

Fv/Fm tests whether or not plant stress affects photosystem II in a dark adapted state. Fv/Fm is the most used chlorophyll fluorescence measuring parameter in the world. "The majority of fluorescence measurements are now made using modulated fluorometers with the leaf poised in a known state." (Neil Baker 2004)

Light that is absorbed by a leaf follows three competitive pathways. It may be used in photochemistry to produce ATP and NADPH used in photosynthesis, it can be re-emitted as fluorescence, or dissipated as heat. The Fv/Fm test is designed to allow the maximum amount of the light energy to take the fluorescence pathway. It compares the dark-adapted leaf pre-photosynthetic fluorescent state, called minimum fluorescence, or Fo, to maximum fluorescence called Fm. In maximum fluorescence, the maximum number of reaction centers have been reduced or closed by a saturating light source. In general, the greater the plant stress, the fewer open reaction centers available, and the Fv/Fm ratio is lowered. Fv/Fm is a measuring protocol that works for many types of plant stress.

In Fv/Fm measurements, after dark adaption, minimum fluorescence is measured, using a modulated light source. This is a measurement of antennae fluorescence using a modulated light intensity that is too low to drive photosynthesis. Next, an intense light flash, or saturation pulse, of a limited duration, is used, to expose the sample, and close all available reaction centers. With all available reaction centers closed, or chemically reduced, maximum fluorescence is measured. The difference between maximum fluorescence and minimum fluorescence is Fv, or variable fluorescence. Fv/Fm is a normalize ratio created by dividing variable fluorescence by maximum fluorescence. It is a measurement ratio that represents the maximum potential quantum efficiency of Photosystem II if all capable reaction centers were open. An Fv/Fm value in the range of 0.79 to 0.84 is the approximate optimal value for many plant species, with lowered values indicating plant stress (Maxwell K., Johnson G. N. 2000), (Kitajima and Butler, 1975). Fv/Fm is a fast test that usually takes a few seconds. It was developed in and around 1975 by Kitajima and Butler. Dark adaptation times vary from about fifteen minutes to overnight. Some researchers will only use pre-dawn values.

Y(II) or ΔF/Fm' and ETR

Y(II) is a measuring protocol that was developed by Bernard Genty with the first publications in 1989 and 1990. It is a light adapted test that allows one to measure plant stress while the plant is undergoing the photosynthetic process at steady-state photosynthesis lighting conditions. Like FvFm, Y(II) represents a measurement ratio of plant efficiency, but in this case, it is an indication of the amount of energy used in photochemistry by photosystem II under steady-state photosynthetic lighting conditions. For most types of plant stress, Y(II) correlates to plant carbon assimilation in a linear fashion in C_4 plants. In C_3 plants, most types of plant stress correlate to carbon assimilation in a curve-linear fashion. According to Maxwell and Johnson, it takes between fifteen and twenty minutes for a plant to reach steady-state photosynthesis at a specific light level. In the field, plants in full sunlight, and not under canopy, or partly cloudy conditions, are considered to be at steady state. In this test, light irradiation levels and leaf temperature must be controlled or measured, because while the Y(II) parameter levels vary with most types of plant stress, it also varies with light level and temperature. Y(II) values will be higher at lower light levels than at higher light levels. Y(II) has the advantage that it is more sensitive to a larger number of plant stress types than Fv/Fm.

ETR, or electron transport rate, is also a light-adapted parameter that is directly related to Y(II) by the equation, ETR = Y(II) × PAR × 0.84 × 0.5. By multiplying Y(II) by the irradiation light level in the PAR range (400 nm to 700 nm) in μmols, multiplied by the average ratio of light absorbed by the leaf 0.84, and multiplied by the average ratio of PSII reaction centers to PSI reaction centers, 0.50, relative ETR measurement is achieved.

Relative ETR values are valuable for stress measurements when comparing one plant to another, as long as the plants to be compared have similar light absorption characteristics. Leaf absorption characteristics can vary by water content, age, and other factors. If absorption differences are a concern, absorption can be measured with the use of an integrating sphere. For more accurate ETR values, the leaf absorption value and the ratio of PSII reaction centers to PSI reaction centers can be included in the equation. If different leaf absorption ratios are an issue, or they are an unwanted variable, then using Y(II) instead of ETR, may be the best choice. Four electrons must be transported for every CO_2molecule assimilated, or O_2 molecule evolved, differences from gas exchange measurements, especially in C_3 plants, can occur under conditions that promote photorespiration, cyclic electron transport, and nitrate reduction. For more detailed information concerning the relationship between fluorescence and gas exchange measurements again refer to Opti-Sciences application note #0509 on Yield measurements.

Quenching Measurements

Quenching measurements have been traditionally used for light stress, and heat stress measurements. In addition, they have been used to study plant photoprotective mechanisms, state transitions, plant photoinhibition, and the distribution of light energy in plants. While they can be used for many types of plant stress measurement, the time required, and the additional expense required for this capability, limit their use. These tests commonly require overnight dark adaptation, and fifteen to twenty minutes in lighted conditions to reach steady state photosynthesis before measurement.

Puddle Model and Lake Model Quenching Parameters

"Understanding of the organization of plant antennae, or plant light collection structures, and reaction centers, where the photosynthetic light reaction actually takes place, has changed over the years. It is now understood that a single antennae does not link only to a single reaction center, as was previously described in the puddle model. Current evidence indicates that reaction centers are connected with shared antennae in terrestrial plants." As a result, the parameters used to provide reliable measurements have changed to represent the newer understanding of this relationship. The model that represents the newer understanding of the antennae - reaction center relationship is called the lake model.

Lake model parameters were provided by Dave Kramer in 2004. Since then, Luke Hendrickson has provided simplified lake model parameters that allow the resurrection of the parameter NPQ, from the puddle model, back into the lake model. This is valuable because there have been so many scientific papers that have used NPQ for plant stress measurement, as compared to papers that have used lake model parameters.

For an in depth overview of quenching, refer to the OSI quenching application note. It discusses all of the parameters used in lake models by Kramer, Hendrickson, and Klughammer. In addition it also reviews puddle model parameters, and quenching relaxation measurements.

OJIP or OJIDP

OJIP or OJIDP is a dark adapted chlorophyll fluorescence technique that is used for plant stress measurement. It has been found that by using a high time resolution scale, the rise to maximum fluorescence from minimum fluorescence has intermediate peaks and dips, designated by the OJID and P nomenclature. Over the years, there have been multiple theories of what the rise, time scale, peaks and dips mean. In addition, there is more than one school as to how this information should be used for plant stress testing (Strasser 2004), (Vredenburg 2004, 2009, 2011). Like Fv/Fm, and the other protocols, the research shows that OJIP works better for some types of plant stress than it does for others.

Choosing the Best Chlorophyll Fluorescence Protocol and Parameter

When choosing the correct protocol, and measuring parameter, for a specific type of plant stress, it is important to understand the limitations of the instrument, and the protocol used. For example, it was found that when measuring Oak leaves, a photosynthesis system could detect heat stress at 30 °C and above, Y(II) could detect heat stress at 35

°C and above, NPQ could detect heat stress at 35 °C and above, and Fv/Fm could only detect heat stress at 45 °C and above. (Haldiman P, & Feller U. 2004) OJIP was found to detect heat stress at 44 °C and above on samples tested. (Strasser 2004)

The relationship between carbon assimilation measurements made by photosynthesis systems of the dark Calvin cycle, and measurements of variable fluorescence of photosystem II (PSII), made by chlorophyll fluorometers of the light reaction, are not always straightforward. For this reason, choosing the correct chlorophyll fluorescence protocol can also be different for C_3 and C_4 plants. It has been found, for example, that Y(II) and ETR are good tests for drought stress in C_4 plants, but a special assay is required to measure drought stress in most C_3 plants at usable levels. In C_3 plants, photorespiration, and the Mehler reaction, are thought to be a principal cause. (Flexas 2000)

Chlorophyll Content Meters

These are instruments that use light transmission through a leaf, at two wavelengths, to determine the greenness and thickness of leaves. Transmission in the infrared range provides a measurement related to leaf thickness, and a wavelength in the red light range is used to determine greenness. The ratio of the transmission of the two wavelengths provides a chlorophyll content index that is referred to as CCI or alternatively as a SPAD index. CCI is a linear scale, and SPAD is a logarithmic scale. These instruments and scales have been shown to correlate to chlorophyll chemical tests for chlorophyll content except at very high levels.

Chlorophyll content meters are commonly used for nutrient plant stress measurement, that includes nitrogen stress, and sulfur stress. Because research has shown, that if used correctly, chlorophyll content meters are reliable for nitrogen management work, these meters are often the instruments of choice for crop fertilizer management because they are relatively inexpensive. Research has demonstrated that by comparing well fertilized plants to test plants, the ratio of the chlorophyll content index of test plants, divided by the chlorophyll content index of well fertilized plants, will provide a ratio that is an indication of when fertilization should occur, and how much should be used. It is common to use a well fertilized stand of crops in a specific field and sometimes in different areas of the same field, as the fertilization reference, due to differences from field to field and within a field. The research done to date uses either ten and thirty measurements on test and well fertilized crops, to provide average values. Studies have been done for corn and wheat. One paper suggests that when the ratio drops below 95%, it is time to fertigate. The amounts of fertilizer are also recommended.

Crop consultants also use these tools for fertilizer recommendations. However, because strict scientific protocols are more time consuming and more expensive, consultants sometimes use well-fertilized plants located in low-lying areas as the standard well-fertilized plants. They typically also use fewer measurements. The evidence for this approach involves anecdotal discussions with crop consultants. Chlorophyll content meters are sensitive to both nitrogen and sulfur stress at usable levels. Chlorophyll fluorometers require a special assay, involving a high actinic light level in combination with nitrogen stress, to measure nitrogen stress at usable levels. In addition, chlorophyll fluorometers will only detect sulfur stress at starvation levels. For best results, chlorophyll content measurements should be made when water deficits are not present.Photosynthesis systems will detect both nitrogen and sulfur stress.

SOIL PH

The **soil pH** is a measure of the acidity or alkalinity in soils. pH is defined as the negative logarithm (base 10) of the activity of hydronium ions (H_+ or, more precisely, H_3O_+aq) in a solution. In water, it normally ranges from -1 to 14, with 7 being neutral. A pH below 7 is acidic and above 7 is alkaline. Soil pH is considered a master variable in soils as it controls many chemical processes that take place. It specifically affects plant nutrient availability by controlling the chemical forms of the nutrient. The optimum pH range for most plants is between 5.5 and 7.0, however many plants have adapted to thrive at pH values outside this range.

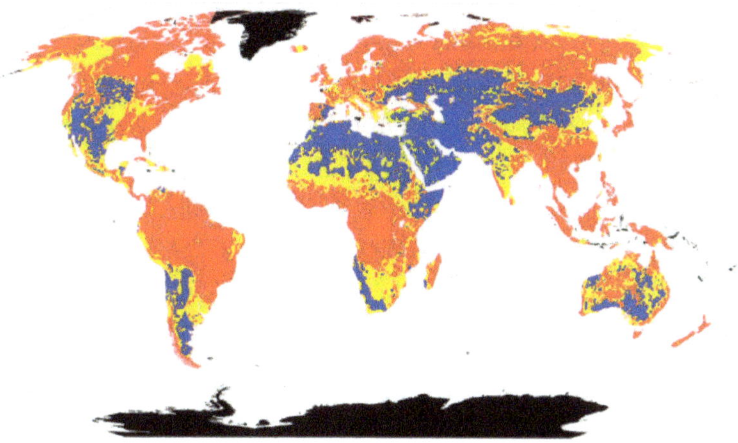

Global variation in soil pH. Red = acidic soil. Yellow = neutral soil. Blue = alkaline soil. Black = no data.

Classification of Soil Ph Ranges

The United States Department of Agriculture Natural Resources Conservation Service, formerly Soil Conservation Service classifies soil pH ranges as follows:

Denomination	pH range
Ultra acidic	< 3.5
Extremely acidic	3.5–4.4
Very strongly acidic	4.5–5.0
Strongly acidic	5.1–5.5
Moderately acidic	5.6–6.0
Slightly acidic	6.1–6.5
Neutral	6.6–7.3
Slightly alkaline	7.4–7.8

Moderately alkaline	7.9–8.4
Strongly alkaline	8.5–9.0
Very strongly alkaline	> 9.0

Sources of Soil Ph

Sources of Acidity

Acidity in soils comes from H^+ and Al^{3+} ions in the soil solution and sorbed to soil surfaces. While pH is the measure of H^+ in solution, Al^{3+} is important in acid soils because between pH 4 and 6, Al^{3+} reacts with water (H_2O) forming $AlOH^{2+}$, and $Al(OH)_2^+$, releasing extra H^+ ions. Every Al^{3+}ion can create 3 H^+ ions. Many other processes contribute to the formation of acid soils including rainfall, fertilizer use, plant root activity and the weathering of primary and secondary soil minerals. Acid soils can also be caused by pollutants such as acid rain and mine spoilings.

- Rainfall: Acid soils are most often found in areas of high rainfall. Excess rainfall leaches base cation from the soil, increasing the percentage of Al^{3+} and H^+ relative to other cations. Additionally, rainwater has a slightly acidic pH of 5.7 due to a reaction with CO_2 in the atmosphere that forms carbonic acid.
- Fertilizer use: Ammonium (NH_4^+) fertilizers react in the soil in a process called nitrification to form nitrate (NO_3^-), and in the process release H^+ ions.
- Plant root activity: Plants take up nutrients in the form of ions (NO_3^-, NH_4^+, Ca^{2+}, $H_2PO_4^-$, etc.), and often, they take up more cations than anions. However plants must maintain a neutral charge in their roots. In order to compensate for the extra positive charge, they will release H^+ ions from the root. Some plants will also exude organic acids into the soil to acidify the zone around their roots to help solubilize metal nutrients that are insoluble at neutral pH, such as iron (Fe).
- Weathering of minerals: Both primary and secondary minerals that compose soil contain Al. As these minerals weather, some components such as Mg, Ca, and K, are taken up by plants, others such as Si are leached from the soil, but due to chemical properties, Fe and Al remain in the soil profile. Highly weathered soils are often characterized by having high concentrations of Fe and Al oxides.
- Acid rain: When atmospheric water reacts with sulfur and nitrogen compounds that result from industrial processes, the result can be the formation of sulfuric and nitric acid in rainwater. However the amount of acidity that is deposited in rainwater is much less, on average, than that created through agricultural activities.
- Mine spoil: Severely acidic conditions can form in soils near mine spoils due to the oxidation of pyrite.
- Potential acid sulfate soils naturally formed in waterlogged coastal and estuarine environments can become highly acidic when drained or excavated.

- Decomposition of organic matter by microorganisms releases CO_2 which when mixed with soil water can form carbonic acid (H_2CO_3).

Sources of Alkalinity

Alkaline soils have a high saturation of base cations (K^+, Ca^{2+}, Mg^{2+} and Na^+). This is due to an accumulation of soluble salts which are classified as either saline soil, sodic soil, saline-sodic soil or alkaline soil. All saline and sodic soils have high salt concentrations, with saline soils being dominated by calcium and magnesium salts and sodic soils being dominated by sodium. Alkaline soils are characterized by the presence of carbonates. Soil in areas with limestone near the surface are alkaline from the calcium carbonate in limestone constantly mixing with the soil.Groundwater sources in these areas contain dissolved limestone.

Effect of Soil Ph on Plant Growth

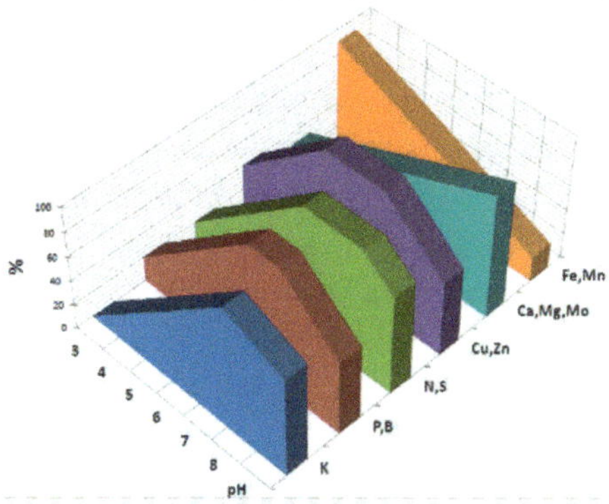

Nutrient availability in relation to soil pH

Acid Affected Soils

Plants grown in acid soils can experience a variety of symptoms including aluminium (Al), hydrogen (H), and/or manganese (Mn) toxicity, as well as nutrient deficiencies of calcium (Ca) and magnesium (Mg).

Aluminium toxicity is the most widespread problem in acid soils. Aluminium is present in all soils, but dissolved Al^{3+} is toxic to plants; Al^{3+} is most soluble at low pH, above pH 5.2 little Al is in soluble form in most soils. Aluminium is not a plant nutrient, and as such, is not actively taken up by the plants, but enters plant roots passively through osmosis. Aluminium inhibits root growth; lateral roots and root tips become thickened and roots lack fine branching; root tips may turn brown. In the root, Al has been shown to interfere with many physiological processes including the uptake and transport of calcium and other essential nutrients, cell division, cell wall formation, and enzyme activity.

Below pH 4, H^+ ions themselves damage root cell membranes.

In soils with high content of manganese-containing minerals, Mn toxicity can become a problem at pH 5.6 and lower. Manganese, like aluminium, becomes increasingly soluble as pH drops, and Mn toxicity symptoms can be seen at pH levels below 5.6. Manganese is an essential plant nutrient, so plants transport Mn into leaves. Classic symptoms of Mn toxicity are crinkling or cupping of leaves.

Nutrient Availability in Relation to Soil Ph

Nutrients needed in large amounts by plants are referred to as macronutrients and include nitrogen (N), phosphorus (P), potassium (K), calcium (Ca), magnesium (Mg) and sulfur (S). Elements that plants need in trace amounts are called trace nutrients or micronutrients. Trace nutrients are not major components of plant tissue but are essential for growth. They include iron (Fe), manganese (Mn), zinc (Zn), copper (Cu), cobalt (Co), molybdenum (Mo), and boron (B). Both macronutrient and micronutrient availability are affected by soil pH. In slightly to moderately alkaline soils, molybdenum and macronutrient (except for phosphorus) availability is increased, but P, Fe, Mn, Zn Cu, and Co levels are reduced and may adversely affect plant growth. In acidic soils, micronutrient availability (except for Mo and Bo) is increased. Nitrogen is supplied as ammonium (NH_4) or nitrate (NO_3) by nitrogen fixation or fertilizer amendments, and dissolved N will have the highest concentrations in soil with pH 6–8. Concentrations of available N are less sensitive to pH than concentration of available P. In order for P to be available for plants, soil pH needs to be in the range 6.0 and 7.5. If pH is lower than 6, P starts forming insoluble compounds with iron (Fe) and aluminium (Al) and if pH is higher than 7.5 P starts forming insoluble compounds with calcium (Ca). Most nutrient deficiencies can be avoided between a pH range of 5.5 to 6.5, provided that soil minerals and organic matter contain the essential nutrients to begin with.

Determining pH

Methods of determining pH include:

- Observation of soil profile: Certain profile characteristics can be indicators of either acid, saline, or sodic conditions. Strongly acidic soils often have poor incorporation of the organic surface layer with the underlying mineral layer. The mineral horizons are distinctively layered in many cases, with a pale eluvial (E) horizon beneath the organic surface; this E is underlain by a darker B horizon in a classic podzol horizon sequence. This is a very rough gauge of acidity as there is no correlation between thickness of the E and soil pH. E horizons a few feet thick in Florida usually have pH just above 5 (merely "strongly acid") while E horizons a few inches thick in New England are "extremely acid" with pH readings of 4.5 or below. In the southern Blue Ridge Mountains, there are ultra acidic soils with pH below 3.5, which have no E horizon. Presence of a caliche layer indicates the presence of calcium carbonates, which are present in alkaline conditions. Also, columnar structure can be an indicator of sodic condition.
- Observation of predominant flora. Calcifuge plants (those that prefer an acidic

soil) include *Erica*, *Rhododendron* and nearly all other Ericaceae species, many birch (*Betula*), foxglove (*Digitalis*), gorse (*Ulex* spp.), and Scots Pine (*Pinus sylvestris*). Calcicole (lime loving) plants include ash trees (*Fraxinus* spp.), honeysuckle (*Lonicera*), *Buddleja*, dogwoods (*Cornus* spp.), lilac (*Syringa*) and *Clematis* species.

- Use of an inexpensive pH testing kit, where in a small sample of soil is mixed with indicator solution which changes colour according to the acidity/alkalinity.
- Use of litmus paper. A small sample of soil is mixed with distilled water, into which a strip of litmus paper is inserted. If the soil is acidic the paper turns red, if alkaline, blue.
- Use of a commercially available electronic pH meter, in which a rod is inserted into moistened soil and measures the concentration of hydrogen ions.

Examples of Plant Ph Preferences

- pH 4.5–5.0: Ericaceae (azalea, bilberry, blueberry, cranberry; heather), hydrangea for blue, (less acidic for pink), liquidambar or sweet gum, orchid, pin oak
- pH 5.0–5.5: Boronia, daphne; Ericaceae: (camellia, heather, rhododendron), ferns, iris, orchids, parsley, conifers (e.g., pine); Poaceae: (maize, millet, rye, oat), radish; Solanales: (potato, sweet potato); Bromeliaceae (pineapple).
- pH 5.5–6.0: Asteraceae: (aster, endive); Brassicaceae: (brussels sprout, kohlrabi), carrot; Cucurbitales: (begonia, chayote or choko); Fabaceae: (bean, crimson clover, peanut, soybean), petunia, rhubarb, violet, most bulbs (canna, daffodil, jonquil), larkspur, primrose
- pH 6.0–6.5 antirrhinum or snapdragon, Brassicaceae: (broccoli, cabbage, candytuft, cauliflower, turnip, wallflower); Cucurbitaceae: (cucumber, pumpkin, squash); Fabaceae: (pea, red clover, white clover), gladiolus, Iceland poppy; Rosales: (cannabis, rose, strawberry); Solanaceae: (eggplant or aubergine, tomato), sweet corn; Violaceae: (pansy, viola), zinnia or zinnea
- pH 6.5–7.0: Amaranthaceae: (beet, spinach); Apiaceae: (celery, parsnip); Asparagales: (asparagus, onion); Asteraceae: (chrysanthemum, dahlia, lettuce), carnation; Fabaceae: (alfalfa, sweet pea), melon, stock, tulip
- pH 7.1–8.0 lilac

Changing Soil Ph

Increasing Ph of Acidic Soil

The most common amendment to increase soil pH is lime ($CaCO_3$ or $MgCO_3$), usually in the form of finely ground agricultural lime. The amount of lime needed to change pH is determined by the mesh size of the lime (how finely it is ground)and the buffering capacity of the soil. A high mesh size (60–100) indicates a finely ground lime, that will react quickly with soil acidity. Buffering capacity of soils is a function of a soils cation exchange capacity, which is in turn determined by the clay content of the soil, the type of clay and the amount of organic matter present. Soils with high clay content, particularly shrink–swell clay, will have a higher buffering capacity than soils with little clay. Soils with high organic matter will also have a higher buffering capacity than those with low organic matter. Soils with high buffering capacity require a greater amount of lime to be added than a soil with a lower buffering capacity for the same incremental change in pH. Other amendments that can be used to increase the pH of soil include wood ash, industrial CaO (burnt lime), and oyster shells. White firewood ash includes metal salts which are important for processes requiring ions such as Na^+ (sodium), K^+ (potassium), Ca^{2+} (calcium), which may or may not be good for the select flora, but decreases the acidic quality of soil. These products increase the pH of soils through the reaction of CO_3^{2-} with H^+ to produce CO_2 and H_2O. Calcium silicate neutralizes active acidity in the soil by removing free hydrogen ions, thereby increasing pH. As its silicate anion captures H^+ ions (raising the pH), it forms monosilicic acid (H_4SiO_4), a neutral solute.

Decreasing Ph of Alkaline Soil

- Iron sulfates or aluminium sulfate as well as elemental sulfur (S) reduce pH through the formation of sulfuric acid.
- Urea, urea phosphate, ammonium nitrate, ammonium phosphates, ammonium sulfate and monopotassium phosphate fertilizers.
- organic matter in the form of plant litter, compost, and manure will decrease soil pH through the decomposition process. Certain acid organic matter such as pine needles, pine sawdust and acid peat are effective at reducing pH.

HEAVY METALS

Heavy metals are generally defined as metals with relatively high densities, atomic weights, or atomic numbers. The criteria used, and whether metalloids are included, vary depending on the author and context. In metallurgy, for example, a heavy metal may be defined on the basis of density, whereas in physics the distinguishing criterion might be atomic number, while a chemist would likely be more concerned with chemical behaviour. More specific definitions have been published, but none of these have been widely accepted. The definitions surveyed in this article encompass up to 96 out of the 118 chemical elements; only mercury, lead and bismuth meet all of them. Despite this lack of agreement, the term (plural or singular) is widely used in science. A density of more than 5 g/cm^3 is sometimes quoted as a commonly used criterion and is used in the body of this article.

The earliest known metals—common metals such as iron, copper, and tin, and precious metals such as silver, gold, and platinum—are heavy metals. From 1809 onwards, light metals, such as magnesium, aluminium, and titanium, were discovered, as well as less well-known heavy metals including gallium, thallium, and hafnium.

Some heavy metals are either essential nutrients (typically iron, cobalt, and zinc), or relatively harmless (such as ruthenium, silver, and indium), but can be toxic in larger amounts or certain forms. Other heavy metals, such as cadmium, mercury, and lead, are highly poisonous. Potential sources of heavy metal poisoning include mining and industrial wastes, agricultural runoff, occupational exposure, and paints and treated timber.

Physical and chemical characterisations of heavy metals need to be treated with caution, as the metals involved are not always consistently defined. As well as being relatively dense, heavy metals tend to be less reactive than lighter metals and have much less soluble sulfides and hydroxides. While it is relatively easy to distinguish a heavy metal such as tungsten from a lighter metal such as sodium, a few heavy metals such as zinc, mercury, and lead have some of the characteristics of lighter metals, and lighter metals such as beryllium, scandium, and titanium have some of the characteristics of heavier metals.

Heavy metals are relatively scarce in the Earth's crust but are present in many aspects of modern life. They are used in, for example, golf clubs, cars, antiseptics, self-cleaning ovens, plastics, solar panels, mobile phones, and particle accelerators.

Crystals of osmium, a heavy metal nearly twice as dense as lead.

Definitions

There is no widely agreed criterion-based definition of a heavy metal. Different meanings may be attached to the term, depending on the context. In metallurgy, for example, a heavy metal may be defined on the basis of density, whereas in physics the distinguishing criterion might be atomic number, and a chemist would likely be more concerned with chemical behaviour.

Density criteria range from above 3.5 g/cm^3 to above 7 g/cm^3.Atomic weight definitions can range from greater than sodium (atomic weight 22.98); greater than 40 (excluding s-

and f-block metals, hence starting with scandium); or more than 200, i.e. from mercury onwards. Atomic numbers of heavy metals are generally given as greater than 20 (calcium); sometimes this is capped at 92 (uranium). Definitions based on atomic number have been criticised for including metals with low densities. For example, rubidium in group (column) 1 of the periodic table has an atomic number of 37 but a density of only 1.532 g/cm^3, which is below the threshold figure used by other authors. The same problem may occur with atomic weight based definitions.

Criteria based on chemical behaviour or periodic table position have been used or suggested. The United States Pharmacopeia includes a test for heavy metals that involves precipitating metallic impurities as their coloured sulfides." In 1997, Stephen Hawkes, a chemistry professor writing in the context of fifty years' experience with the term, said it applied to "metals with insoluble sulfides and hydroxides, whose salts produce colored solutions in water and whose complexes are usually colored". On the basis of the metals he had seen referred to as heavy metals, he suggested it would useful to define them as (in general) all the metals in periodic table columns 3 to 16 that are in row 4 or greater, in other words, the transition metals and post-transition metals. The lanthanides satisfy Hawkes' three-part description; the status of the actinides is not completely settled.

In biochemistry, heavy metals are sometimes defined—on the basis of the Lewis acid (electronic pair acceptor) behaviour of their ions in aqueous solution—as class B and borderline metals. In this scheme, class A metal ions prefer oxygen donors; class B ions prefer nitrogen or sulfur donors; and borderline or ambivalent ions show either class A or B characteristics, depending on the circumstances. Class A metals, which tend to have low electronegativity and form bonds with large ionic character, are the alkali and alkaline earths, aluminium, the group 3 metals, and the lanthanides and actinides. Class B metals, which tend to have higher electronegativity and form bonds with considerable covalent character, are mainly the heavier transition and post-transition metals. Borderline metals largely comprise the lighter transition and post-transition metals (plus arsenic and antimony). The distinction between the class A metals and the other two categories is sharp. A frequently cited proposal to use these classification categories instead of the more evocative name *heavy metal* has not been widely adopted.

List of Heavy Metals Based on Density

A density of more than 5 g/cm^3 is sometimes mentioned as a common heavy metal defining factor and, in the absence of a unanimous definition, is used to populate this list and (unless otherwise stated) guide the remainder of the article. Metalloids meeting the applicable criteria–arsenic and antimony for example—are sometimes counted as heavy metals, particularly in environmental chemistry, as is the case here. Selenium (density 4.8 g/cm^3) is also included in the list. It falls marginally short of the density criterion and is less commonly recognised as a metalloid but has a waterborne chemistry similar in some respects to that of arsenic and antimony. Other metals sometimes classified or treated as "heavy" metals, such as beryllium (density 1.8 gm/cm^3), aluminium (2.7 gm/cm^3), calcium (1.55 gm/cm^3), and barium (3.6 gm/cm^3) are here treated as light metals and, in general, are not further considered.

Produced mainly by commercial mining (informally classified by economic significance)				
Strategic (30)  Considered vital to multiple nations' strategic interests[57] *These 30 include 22 listed here and 8 below (6 precious & 2 commodity).*	Antimony† Cerium Dysprosiur Erbium Europium Gadolinium	Gallium Germaniun Holmium Indium Lanthanum Lutetium	Neodymiur Niobium Praseodyn Samarium ☢ Tantalum Terbium	Thulium Tungsten Uranium ☢ Ytterbium
Precious (8) 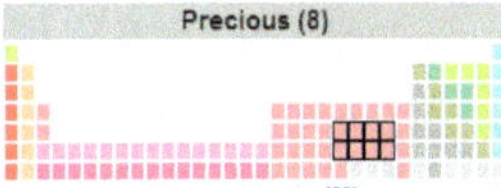 Rare and costly[58]	*Strategic:* Iridium Osmium	Palladium Platinum	Rhodium Rutheniur	*Non-strategic:* Gold Silver
Commodity (9) 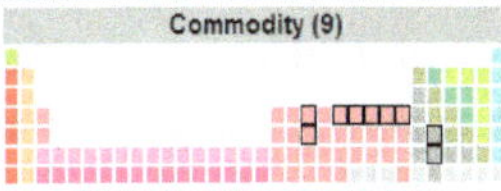 Traded by the tonne on the LME	*Strategic:* Chromium Cobalt	*Non-strategic:* Copper Iron Lead	Molybdenu Nickel Tin	Zinc
Minor (14) 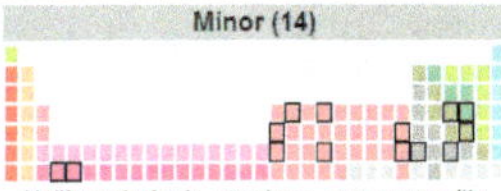 Neither strategic, precious, nor commodity	Arsenic† Bismuth Cadmium Hafnium	Manganese Mercury Protactiniu ☢	Rhenium Selenium‡ Tellurium† Thallium	Thorium ☢ Vanadiur Zirconium
Produced mainly by artificial transmutation (informally classified by stability)				
Long-lived (15) 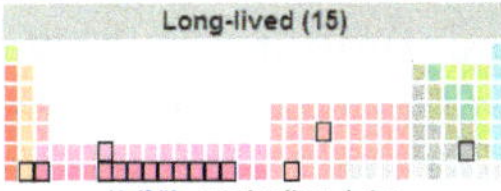 Half-life greater than 1 day	Actinium ☢¶ Americium ☢ Berkelium ☢ Californiur ☢	Curium☢ Dubnium ☢ Einsteiniur ☢ Fermium ☢	Mendeleviu ☢ Neptunium ☢¶ Plutonium ☢¶ Polonium ☢¶	Promethiu ☢¶ Radium☢ ¶ Technetiu ☢¶
Ephemeral (16) 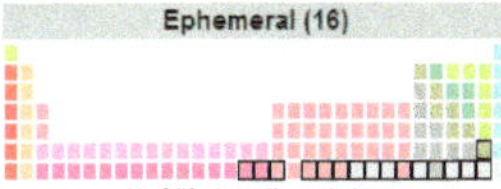 Half-life less than 1 day	Astatine ☢‡¶ Bohrium ☢ Coperniciu ☢ Darmstadti ☢	Flerovium ☢ Hassium ☢ Lawrenciur ☢ Livermoriur ☢	Meitnerium ☢ Moscovium ☢ Nihonium ☢ Nobelium ☢	Roentgeni ☢ Rutherfor ☢ Seaborgiu ☢ Tennessir ☢

†Antimony, arsenic, germanium and tellurium are commonly recognised as metalloids; selenium less commonly so.

‡Astatine is predicted to be a metal.

☢ All isotopes of these 34 elements are unstable and hence radioactive. While this is also true of bismuth, it is not so marked since its half-life of 19 billion billion years is over a billion times the 13.8 billion year estimated age of the universe.

¶These eight elements do occur naturally but in amounts too small for economically viable extraction.

Origins and Use of the Term

The heaviness of naturally occurring metals such as gold, copper, and iron may have been noticed in prehistory and, in light of their malleability, led to the first attempts to craft metal ornaments, tools, and weapons. All metals discovered from then until 1809 had relatively high densities; their heaviness was regarded as a singularly distinguishing criterion.

From 1809 onwards, light metals such as sodium, potassium, and strontium were isolated. Their low densities challenged conventional wisdom and it was proposed to refer to them as *metalloids* (meaning "resembling metals in form or appearance"). This suggestion was ignored; the new elements came to be recognised as metals, and the term metalloid was then used to refer to nonmetallic elements and, later, elements that were hard to describe as either metals or nonmetals.

An early use of the term "heavy metal" dates from 1817, when the German chemist Leopold Gmelin divided the elements into nonmetals, light metals, and heavy metals. Light metals had densities of 0.860–5.0 g/cm^3; heavy metals 5.308–22.000. The term later became associated with elements of high atomic weight or high atomic number. It is sometimes used interchangeably with the term *heavy element*. For example, in discussing the history of nuclear chemistry, Magee notes that the actinides were once thought to represent a new heavy element transition group whereas Seaborg and co-workers, "favoured ... a heavy metal rare-earth like series ...». In astronomy, however, a heavy element is any element heavier than hydrogen and helium.

Criticism

In 2002, Scottish toxicologist John Duffus reviewed the definitions used over the previous 60 years and concluded they were so diverse as to effectively render the term meaningless. Along with this finding, the heavy metal status of some metals is occasionally challenged on the grounds that they are too light, or are involved in biological processes, or rarely constitute environmental hazards. Examples include scandium (too light); vanadium to zinc (biological processes); and rhodium, indium, and osmium (too rare).

Popularity

Despite its questionable meaning, references to the term *heavy metal* appear regularly in scientific literature. A 2010 study found that it had been increasingly used and seemed to have become part of the language of science. It is said to be an acceptable term, given its convenience and familiarity, as long as it is accompanied by a strict definition. The counterparts to the heavy metals, the *light metals*, are alluded to by The Minerals, Metals and Materials Society as including "aluminium, magnesium, beryllium, titanium, lithium, and other reactive metals." The named metals have densities of 0.534 to 4.54 g/cm^3.

Biological Role

Trace amounts of some heavy metals, mostly in period 4, are required for certain biological processes. These are iron and copper (oxygen and electron transport); cobalt (complex syntheses and cell metabolism); zinc (hydroxylation); vanadium and manganese (enzyme regulation or functioning); chromium (glucose utilisation); nickel (cell growth); arsenic (metabolic growth in some animals and possibly in humans) and selenium (antioxidant functioning and hormone production). Periods 5 and 6 contain fewer essential heavy metals, consistent with the general pattern that heavier elements tend to be less abundant and that scarcer elements are less likely to be nutritionally essential. In period 5, molybdenum is required for the catalysis of redox reactions; cadmium is used by some marine diatoms for the same purpose; and tin may be required for growth in a few species. In period 6, tungsten is required by some bacteria for metabolic processes. An average 70 kg human body is about 0.01% heavy metals (~7 g, equivalent to the weight of two dried peas, with iron at 4 g, zinc at 2.5 g, and lead at 0.12 g comprising the three main constituents), 2% light metals (~1.4 kg, the weight of a bottle of wine) and nearly 98% nonmetals (mostly water).

A deficiency of any of these period 4–6 essential heavy metals may increase susceptibility to heavy metal poisoning. A few non-essential heavy metals have also been observed to have biological effects. Gallium, germanium (a metalloid), indium, and most lanthanides can stimulate metabolism, and titanium promotes growth in plants (though it is not always considered a heavy metal).

Toxicity

The focus of this section is mainly on the more serious toxic effects of heavy metals, including cancer, brain damage or death, rather than the harm they may cause to one more of the skin, lungs, stomach, kidneys, liver, or heart. For more specific information see Metal toxicity, Toxic heavy metal, or the articles on individual elements or compounds.

Heavy metals are often assumed to be highly toxic or damaging to the environment. Some are, while certain others are toxic only if taken in excess or encountered in certain forms.

Environmental Heavy Metals

Chromium, arsenic, cadmium, mercury, and lead have the greatest potential to cause harm on account of their extensive use, the toxicity of some of their combined or elemental forms, and their widespread distribution in the environment. Hexavalent chromium, for example, is highly toxic as are mercury vapour and many mercury compounds. These five elements have a strong affinity for sulfur; in the human body they usually bind, via thiol groups (–SH), to enzymes responsible for controlling the speed of metabolic reactions. The resulting sulfur-metal bonds inhibit the proper functioning of the enzymes involved; human health deteriorates, sometimes fatally. Chromium (in its hexavalent form) and arsenic are carcinogens; cadmium causes a degenerative bone disease; and mercury and lead damage the central nervous system.

Chromium crystals and 1 cm^3 cube

Arsenic, sealed in a container to stop tarnishing

Cadmium bar and 1 cm^3 cube

Mercury being poured into a petri dish

Oxidised lead nodules and 1 cm^3 cube

Lead is the most prevalent heavy metal contaminant. Levels in the aquatic environments of industrialised societies have been estimated to be two to three times those of pre-industrial levels. As a component of tetraethyl lead, $(CH_3CH_2)_4Pb$, it was used extensively in gasoline during the 1930s–1970s. Although the use of leaded gasoline was largely phased out in North America by 1996, soils next to roads built before this time retain high lead concentrations. Later research demonstrated a statistically significant correlation between the usage rate of leaded gasoline and violent crime in the United States; taking into account a 22-year time lag (for the average age of violent criminals), the violent crime curve virtually tracked the lead exposure curve.

Other heavy metals noted for their potentially hazardous nature, usually as toxic environmental pollutants, include manganese (central nervous system damage); cobalt and nickel (carcinogens); copper, zinc, selenium and silver (endocrine disruption, congenital disorders, or general toxic effects in fish, plants, birds, or other aquatic organisms); tin, as organotin (central nervous system damage);antimony (a suspected carcinogen); and thallium (central nervous system damage).

Nutritionally Essential Heavy Metals

Heavy metals essential for life can be toxic if taken in excess; some have notably toxic forms. Vanadium pentoxide (V_2O_5) is carcinogenic in animals and, when inhaled, causes DNA damage. The purple permanganate ion $MnO–_4$ is a liver and kidney poison. Ingesting more than 0.5 grams of iron can induce cardiac collapse; such overdoses most commonly occur in children and may result in death within 24 hours.Nickel carbonyl ($Ni_2(CO)_4$), at 30 parts per million, can cause respiratory failure, brain damage and death. Imbibing a gram or more of copper sulfate ($Cu(SO_4)_2$) can be fatal; survivors may be left with major organ damage. More than five milligrams of selenium is highly toxic; this is roughly ten times the 0.45 milligram recommended maximum daily intake; long-term poisoning can have paralytic effects.

Other Heavy Metals

A few other non-essential heavy metals have one or more toxic forms. Kidney failure and fatalities have been recorded arising from the ingestion of germanium dietary supplements (~15 to 300 g in total consumed over a period of two months to three years). Exposure to osmium tetroxide (OsO_4) may cause permanent eye damage and can lead to respiratory failure and death. Indium salts are toxic if more than few milligrams are ingested and will affect the kidneys, liver, and heart. Cisplatin ($PtCl_2(NH_3)_2$), which is an important drug used to kill cancer cells, is also a kidney and nerve poison. Bismuth compounds can cause liver damage if taken in excess; insoluble uranium compounds, as well as the dangerous radiation they emit, can cause permanent kidney damage.

Exposure Sources

Heavy metals can degrade air, water, and soil quality, and subsequently cause health issues in plants, animals, and people, when they become concentrated as a result of industrial activities. Common sources of heavy metals in this context include mining and industrial wastes; vehicle emissions; lead-acid batteries; fertilisers; paints; and treated timber; aging water supply infrastructure; and microplastics floating in the world's oceans. Recent examples of heavy metal contamination and health risks include the occurrence of Minamata disease, in Japan (1932–1968; lawsuits ongoing as of 2016); the Bento Rodrigues dam disaster in Brazil, and high levels of lead in drinking water supplied to the residents of Flint, Michigan, in the north-east of the United States.

Formation, Abundance, Occurrence, and Extraction

Heavy metals up to the vicinity of iron (in the periodic table) are largely made via stellar nucleosynthesis. In this process, lighter elements from hydrogen to silicon undergo successive fusion reactions inside stars, releasing light and heat and forming heavier elements with higher atomic numbers.

Heavier heavy metals are not usually formed this way since fusion reactions involving such nuclei would consume rather than release energy. Rather, they are largely synthesised (from elements with a lower atomic number) by neutron capture, with the two main modes of this repetitive capture being the s-process and the r-process. In the s-process ("s" stands for "slow"), singular captures are separated by years or decades, allowing the less stable nuclei to beta decay, while in the r-process ("rapid"), captures happen faster than nuclei can decay. Therefore, the s-process takes a more or less clear path: for example, stable cadmium-110 nuclei are successively bombarded by free neutrons inside a star until they form cadmium-115 nuclei which are unstable and decay to form indium-115 (which is nearly stable, with a half-life 30000 times the age of the universe). These nuclei capture neutrons and form indium-116, which is unstable, and decays to form tin-116, and so on. In contrast, there is no such path in the r-process. The s-process stops at bismuth due to the short half-lives of the next two elements, polonium, and astatine, which decay to bismuth or lead. The r-process is so fast it can skip this zone of instability and go on to create heavier elements such as thorium and uranium.

Heavy metals condense in planets as a result of stellar evolution and destruction processes. Stars lose much of their mass when it is ejected late in their lifetimes, and sometimes thereafter as a result of a neutron star merger, thereby increasing the abundance of elements heavier than helium in the interstellar medium. When gravitational attraction causes this matter to coalesce and collapse new stars and planets are formed.

The Earth's crust is made of approximately 5% of heavy metals by weight, with iron comprising 95% of this quantity. Light metals (~20%) and nonmetals (~75%) make up the other 95% of the crust. Despite their overall scarcity, heavy metals can become concentrated in economically extractable quantities as a result of mountain building, erosion, or other geological processes.

Heavy metals are primarily found as lithophiles (rock-loving) or chalcophiles (ore-loving). Lithophile heavy metals are mainly f-block elements and the more reactive of the d-block elements. They have a strong affinity for oxygen and mostly exist as relatively low density silicate minerals.Chalcophile heavy metals are mainly the less reactive d-block elements, and period 4–6 p-block metals and metalloids. They are usually found in (insoluble) sulfide minerals. Being denser than the lithophiles, hence sinking lower into the crust at the time of its solidification, the chalcophiles tend to be less abundant than the lithophiles.

On the other hand, gold is a siderophile, or iron-loving element. It does not readily form compounds with either oxygen or sulfur. At the time of the Earth's formation, and as the most noble (inert) of metals, gold sank into the core due to its tendency to form high-density metallic alloys. Consequently, it is a relatively rare metal. Some other (less) noble heavy metals—molybdenum, rhenium, the platinum group metals (ruthenium, rhodium, palladium, osmium, iridium, and platinum), germanium, and tin—can be counted as siderophiles but only in terms of their primary occurrence in the Earth (core, mantle and crust), rather the crust. These metals otherwise occur in the crust, in small quantities, chiefly as chalcophiles (less so in their native form).

Concentrations of heavy metals below the crust are generally higher, with most being found in the largely iron-silicon-nickel core. Platinum, for example, comprises approximately 1 part per billion of the crust whereas its concentration in the core is thought to be nearly 6,000 times higher. Recent speculation suggests that uranium (and thorium) in the core may generate a substantial amount of the heat that drives plate tectonics and (ultimately) sustains the Earth's magnetic field.[n 23]

The winning of heavy metals from their ores is a complex function of ore type, the chemical properties of the metals involved, and the economics of various extraction methods. Different countries and refineries may use different processes, including those that differ from the brief outlines listed here.

Broadly speaking, and with some exceptions, lithophile heavy metals can be extracted from their ores by electrical or chemical treatments, while chalcophile heavy metals are obtained by roasting their sulphide ores to yield the corresponding oxides, and then heating these to obtain the raw metals. Radium occurs in quantities too small to be economically mined and is instead obtained from spent nuclear fuels. The chalcophile platinum group metals (PGM) mainly occur in small (mixed) quantities with other chalcophile ores. The ores involved need to be smelted, roasted, and then leached with sulfuric acid to produce a residue of PGM. This is chemically refined to obtain the individual metals in their pure forms. Compared to other metals, PGM are expensive due to their scarcity and high production costs.

Gold, a siderophile, is most commonly recovered by dissolving the ores in which it is found in a cyanide solution. The gold forms a dicyanoaurate(I), for example: 2 Au + H_2O +½ O_2 + KCN → 2 K[Au$(CN)_2$] + 2 KOH. Zinc is added to the mix and, being more reactive than gold, displaces the gold: 2[Au$(CN)_2$] + Zn → K_2[Zn$(CN)_4$] + 2 Au. The gold precipitates out of solution as a sludge, and is filtered off and melted.

Properties Compared with Light Metals

Some general physical and chemical properties of light and heavy metals are summarised in the table. The comparison should be treated with caution since the terms light metal and heavy metal are not always consistently defined. Also the physical properties of hardness and tensile strength can vary widely depending on purity, grain size and pre-treatment.

Properties of light and heavy metals

Physical properties	**Light metals**	**Heavy metals**
Density	Usually lower	Usually higher
Hardness	Tend to be soft, easily cut or bent	Most are quite hard
Thermal expansivity	Mostly higher	Mostly lower
Melting point	Mostly low	Low to very high
Tensile strength	Mostly lower	Mostly higher
Chemical properties	**Light metals**	**Heavy metals**
Periodic table location	Most found in groups 1 and 2	Nearly all found in groups 3 through 16
Abundance in Earth's crust	More abundant	Less abundant
Main occurrence (or source)	Lithophiles	Lithophiles or chalcophiles (Au is a siderophile)
Reactivity	More reactive	Less reactive
Sulfides	Soluble to insoluble[n 25]	Extremely insoluble
Hydroxides	Soluble to insoluble[n 26]	Generally insoluble
Salts	Mostly form colourless solutions in water	Mostly form coloured solutions in water
Complexes	Mostly colourless	Mostly coloured
Biological role	Include macronutrients (Na, Mg, K, Ca)	Include micronutrients (V, Cr, Mn, Fe, Co, Ni, Cu, Zn, Mo)

These properties make it relatively easy to distinguish a light metal like sodium from a heavy metal like tungsten, but the differences become less clear at the boundaries. Light structural metals like beryllium, scandium, and titanium have some of the characteristics of heavy metals, such as higher melting points;[n 27] post-transition heavy metals like zinc, cadmium, and lead have some of the characteristics of light metals, such as being relatively soft, and having lower melting points; and forming mainly colourless complexes.

Uses

Heavy metals are present in nearly all aspects of modern life. Iron may be the most common as it accounts for 90% of all refined metals. Platinum may be the most ubiquitous given it is said to be found in, or used to produce, 20% of all consumer goods.

Some common uses of heavy metals depend on the general characteristics of metals such as electrical conductivity and reflectivity or the general characteristics of heavy metals such as density, strength, and durability. Other uses depend on the characteristics of the specific element, such as their biological role as nutrients or poisons or some other specific atomic properties. Examples of such atomic properties include: partly filled d- or f- orbitals (in many of the transition, lanthanide, and actinide heavy metals) that enable the formation of coloured compounds; the capacity of most heavy metal ions (such as platinum, cerium or bismuth) to exist in different oxidation states and therefore act as catalysts;poorly overlapping 3d or 4f orbitals (in iron, cobalt, and nickel, or the lanthanide heavy metals from europium through thulium) that give rise to magnetic effects; and high atomic numbers and electron densities that underpin their nuclear science applications. Typical uses of heavy metals can be broadly grouped into the following six categories.

Weight- or Density-Based

In a cello (example shown above) or a viola the C-string sometimes incorporates tungsten; its high density permits a smaller diameter string and improves responsiveness.

Some uses of heavy metals, including in sport, mechanical engineering, military ordnance, and nuclear science, take advantage of their relatively high densities. In underwater diving, lead is used as a ballast; in handicap horse racing each horse must carry a specified lead weight, based on factors including past performance, so as to equalize the chances of the various competitors. In golf, tungsten, brass, or copper inserts in fairway clubs and irons lower the centre of gravity of the club making it easier to get the ball into the air; and golf balls with tungsten cores are claimed to have better flight characteristics. In fly fishing, sinking fly lines have a PVC coating embedded with tungsten powder, so that they sink at the required rate. In track and field sport, steel balls used in the hammer throw and shot put events are filled with lead in order to attain the minimum weight required under international rules. Tungsten was used in hammer throw balls at least up to 1980; the minimum size of the ball was increased in 1981 to eliminate the need for what was, at that time, an expensive metal (triple the cost of other hammers) not generally available in all countries. Tungsten hammers were so dense that they penetrated too deeply into the turf.

In mechanical engineering, heavy metals are used for ballast in boats, aeroplanes, and motor vehicles; or in balance weights on wheels and crankshafts, gyroscopes, and propellers, and centrifugal clutches, in situations requiring maximum weight in minimum space (for example in watch movements).

In military ordnance, tungsten or uranium is used in armour plating and armour piercing projectiles, as well as in nuclear weapons to increase efficiency (by reflecting neutrons and momentarily delaying the expansion of reacting materials). In the 1970s, tantalum was found to be more effective than copper in shaped charge and explosively formed anti-armour weapons on account of its higher density, allowing greater force concentration, and better deformability. Less-toxic heavy metals, such as copper, tin, tungsten, and bismuth, and probably manganese (as well as boron, a metalloid), have replaced lead and antimony in the green bullets used by some armies and in some recreational shooting munitions. Doubts have been raised about the safety (or green credentials) of tungsten.

Because denser materials absorb more radioactive emissions than lighter ones, heavy metals are useful for radiation shielding and to focus radiation beams in linear accelerators and radiotherapy applications.

Strength- or Durability-Based

The strength or durability of heavy metals such as chromium, iron, nickel, copper, zinc, molybdenum, tin, tungsten, and lead, as well as their alloys, makes them useful for the manufacture of artefacts such as tools, machinery, appliances, utensils, pipes, railroad tracks, buildings and bridges, automobiles, locks, furniture, ships, planes, coinage and jewellery.They are also used as alloying additives for enhancing the properties of other metals. Of the two dozen elements that have been used in the world's monetised coinage only two, carbon and aluminium, are not heavy metals. Gold, silver, and platinum are used in jewelleryas are (for example) nickel, copper, indium, and cobalt in coloured gold. Low-cost jewellery and children's toys may be made, to a significant degree, of heavy metals such as chromium, nickel, cadmium, or lead.

The Statue of Liberty. A stainless steel alloy armature provides structural strength; a copper skin confers corrosion resistance.

Copper, zinc, tin, and lead are mechanically weaker metals but have useful corrosion prevention properties. While each of them will react with air, the resulting patinas of either various copper salts, zinc carbonate, tin oxide, or a mixture of lead oxide, carbonate, and sulfate, confer valuable protective properties. Copper and lead are therefore used, for example, as roofing materials; zinc acts as an anti-corrosion agent in galvanised steel; and tin serves a similar purpose on steel cans.

The workability and corrosion resistance of iron and chromium are increased by adding gadolinium; the creep resistance of nickel is improved with the addition of thorium. Tellurium is added to copper and steel alloys to improve their machinability; and to lead to make it harder and more acid-resistant.

Biological and Chemical

Cerium(IV) oxide (sample shown above) is used as a catalyst in self-cleaning ovens.

The biocidal effects of some heavy metals have been known since antiquity. Platinum, osmium, copper, ruthenium, and other heavy metals, including arsenic, are used in anti-cancer treatments, or have shown potential. Antimony (anti-protozoal), bismuth (anti-ulcer), gold (anti-arthritic), and iron (anti-malarial) are also important in medicine. Copper, zinc, silver, gold, or mercury are used in antiseptic formulations;small amounts of some heavy metals are used to control algal growth in, for example, cooling towers.Depending on their intended use as fertilisers or biocides, agrochemicals may contain heavy metals such as chromium, cobalt, nickel, copper, zinc, arsenic, cadmium, mercury, or lead.

Selected heavy metals are used as catalysts in fuel processing (rhenium, for example), synthetic rubber and fibre production (bismuth), emission control devices (palladium), and in self-cleaning ovens (where cerium(IV) oxide in the walls of such ovens helps oxidise carbon-based cooking residues). In soap chemistry, heavy metals form insoluble soaps that are used in lubricating greases, paint dryers, and fungicides (apart from lithium, the alkali metals and the ammonium ion form soluble soaps).

Colouring and Optics

Neodymium sulfate ($Nd_2(SO_4)_3$), used to colour glassware

The colours of glass, ceramic glazes, paints, pigments, and plastics are commonly produced by the inclusion of heavy metals (or their compounds) such as chromium, manganese, cobalt, copper, zinc, selenium, zirconium, molybdenum, silver, tin, praseodymium, neodymium, erbium, tungsten, iridium, gold, lead, or uranium. Tattoo inks may contain heavy metals, such as chromium, cobalt, nickel, and copper. The high reflectivity of some heavy metals is important in the construction of mirrors, including precision astronomical instruments. Headlight reflectors rely on the excellent reflectivity of a thin film of rhodium.

Electronics, Magnets, and Lighting

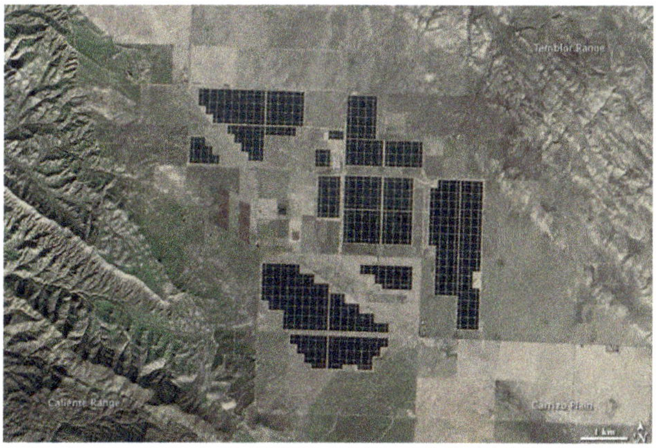

The Topaz Solar Farm, in southern California, features 9 million cadmium-tellurium photovoltaic modules covering an area of 25.6 square kilometres (9.5 square miles).

Heavy metals or their compounds can be found in electronic components, electrodes, and wiring and solar panels where they may be used as either conductors, semiconductors, or insulators. Molybdenum powder is used in circuit board inks. Ruthenium(IV) oxide coated titanium anodes are used for the industrial production of chlorine. Home electrical systems, for the most part, are wired with copper wire for its good conducting properties. Silver and gold are used in electrical and electronic devices, particularly in contact switches, as a result of their high electrical conductivity and capacity to resist or minimise the formation of impurities on their surfaces. The semiconductors cadmium telluride and gallium arsenide are used to make solar panels. Hafnium oxide, an insulator, is used as a voltage controller in microchips; tantalum oxide, another insulator, is used in capacitors in mobile phones. Heavy metals have been used in batteries for over 200 years, at least since Volta invented his copper and silver voltaic pile in 1800. Promethium, lanthanum, and mercury are further examples found in, respectively, atomic, nickel-metal hydride, and button cell batteries.

Magnets are made of heavy metals such as manganese, iron, cobalt, nickel, niobium, bismuth, praseodymium, neodymium, gadolinium, and dysprosium. Neodymium magnets are the strongest type of permanent magnet commercially available. They are key components of, for example, car door locks, starter motors, fuel pumps, and power windows.

Heavy metals are used in lighting, lasers, and light-emitting diodes (LEDs). Flat panel displays incorporate a thin film of electrically conducting indium tin oxide. Fluorescent lighting relies on mercury vapour for its operation. Ruby lasers generate deep red beams by exciting chromium atoms; the lanthanides are also extensively employed in lasers. Gallium, indium, and arsenic; and copper, iridium, and platinum are used in LEDs (the latter three in organic LEDs).

Nuclear

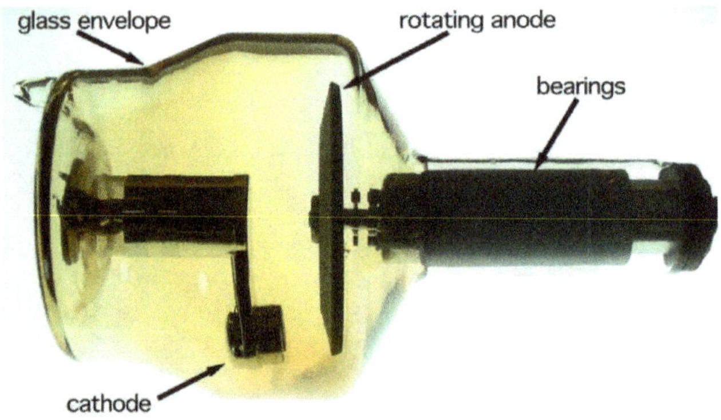

An X-ray tube with a rotating anode, typically a tungsten-rhenium alloy on a molybdenum core, backed with graphite

Niche uses of heavy metals with high atomic numbers occur in diagnostic imaging, electron microscopy, and nuclear science. In diagnostic imaging, heavy metals such as cobalt or tungsten make up the anode materials found in x-ray tubes. In electron microscopy, heavy metals such as lead, gold, palladium, platinum, or uranium are used to make conductive coatings and to introduce electron density into biological specimens by staining, negative staining, or vacuum deposition. In nuclear science, nuclei of heavy metals such as chromium, iron, or zinc are sometimes fired at other heavy metal targets to produce superheavy elements; heavy metals are also employed as spallation targets for the production of neutrons or radioisotopes such as astatine (using lead, bismuth, thorium, or uranium in the latter case).

TOXIC HEAVY METAL

A toxic heavy metal is any relatively dense metal or metalloid that is noted for its potential toxicity, especially in environmental contexts. The term has particular application to cadmium, mercury, lead and arsenic, all of which appear in the World Health Organisation's list of 10 chemicals of major public concern. Other examples include manganese, chromium, cobalt, nickel, copper, zinc, selenium, silver, antimony and thallium.

Heavy metals are found naturally in the earth. They become concentrated as a result of human caused activities and can enter plant, animal, and human tissues via inhalation, diet, and manual handling. Then, they can bind to and interfere with the functioning of vital cellular components. The toxic effects of arsenic, mercury, and lead were known to the ancients, but methodical studies of the toxicity of some heavy metals appear to date from only 1868. In humans, heavy metal poisoning is generally treated by the administration of chelating agents. Some elements otherwise regarded as toxic heavy metals are essential, in small quantities, for human health.

A 25-foot (7.6 m) wall of coal fly ash contaminated with toxic heavy metals, resulting from the release of 5.4 million cubic yards of coal fly ash slurry into the Emory River, Tennessee, and nearby land and water features, in December 2008.Testing showed significantly elevated levels of arsenic, copper, barium, cadmium, chromium, lead, mercury, nickel, and thallium in samples of slurry and river water. Cleanup costs may exceed $1.2 billion.

Contamination Sources

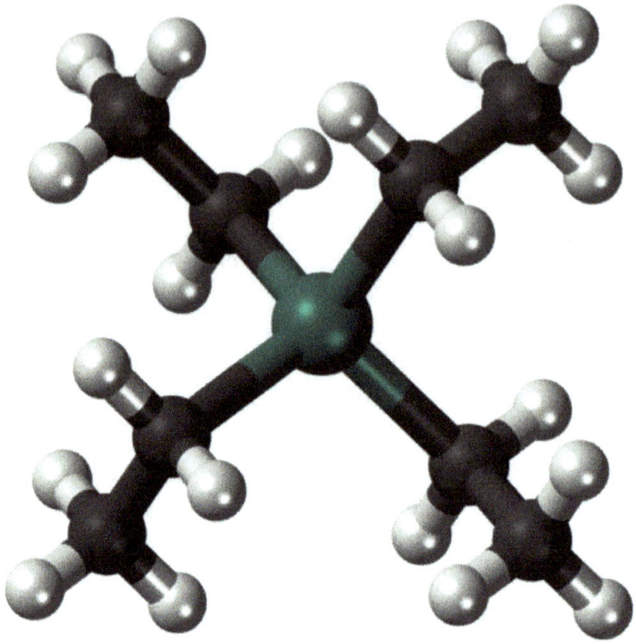

Tetraethyl lead is one of the most significant heavy metal contaminants in recent use.

Heavy metals are found naturally in the earth, and become concentrated as a result of human caused activities. Common sources are from mining and industrial wastes; vehicle emissions; lead-acid batteries; fertilisers; paints; treated woods; aging water supply infrastructure; and microplastics floating in the world's oceans. Arsenic, cadmium and lead may be present in children's toys at levels that exceed regulatory standards. Lead can be used in toys as a stabilizer, color enhancer, or anti-corrosive agent. Cadmium is sometimes employed as a stabilizer, or to increase the mass and luster of toy jewelry. Arsenic is thought to be used in connection with coloring dyes. Regular imbibers of illegally distilled alcohol may be exposed to arsenic or lead poisoning the source of which is arsenic-contaminated lead used to solder the distilling apparatus. Rat poison used in grain and mash stores may be another source of the arsenic.

Lead is the most prevalent heavy metal contaminant. As a component of tetraethyl lead, (CH 3CH_2)Pb, it was used extensively in gasoline during the 1930s–1970s. Lead levels in the aquatic environments of industrialised societies have been estimated to be two to three times those of pre-industrial levels.Although the use of leaded gasoline was largely phased out in North America by 1996, soils next to roads built before this time retain high lead concentrations. Lead (from lead azide or lead styphnate used in firearms) gradually accumulates at firearms training grounds, contaminating the local environment and exposing range employees to a risk of lead poisoning.

Entry Routes

Heavy metals enter plant, animal and human tissues via air inhalation, diet and manual handling. Motor vehicle emissions are a major source of airborne contaminants including arsenic, cadmium, cobalt, nickel, lead, antimony, vanadium, zinc, platinum, palladium and rhodium. Water sources (groundwater, lakes, streams and rivers) can be polluted by heavy metals leaching from industrial and consumer waste; acid rain can exacerbate this process by releasing heavy metals trapped in soils. Plants are exposed to heavy metals through the uptake of water; animals eat these plants; ingestion of plant- and animal-based foods are the largest sources of heavy metals in humans. Absorption through skin contact, for example from contact with soil, is another potential source of heavy metal contamination. Toxic heavy metals can bioaccumulate in organisms as they are hard to metabolize.

Detrimental Effects

Heavy metals "can bind to vital cellular components, such as structural proteins, enzymes, and nucleic acids, and interfere with their functioning." Symptoms and effects can vary according to the metal or metal compound, and the dose involved. Broadly, long-term exposure to toxic heavy metals can have carcinogenic, central and peripheral nervous system and circulatory effects. For humans, typical presentations associated with exposure to any of the "classical" toxic heavy metals, or chromium (another toxic heavy metal) or arsenic (a metalloid), are shown in the table.

Element	**Acute exposure** *usually a day or less*	**Chronic exposure** *often months or years*
Cadmium	Pneumonitis (lung inflammation)	Lung cancer Osteomalacia (softening of bones) Proteinuria (excess protein in urine; possible kidney damage)
Mercury	Diarrhea Fever Vomiting	Stomatitis (inflammation of gums and mouth) Nausea Nephrotic syndrome (nonspecific kidney disorder) Neurasthenia (neurotic disorder) Parageusia (metallic taste) Pink Disease (pain and pink discoloration of hands and feet) Tremor
Lead	Encephalopathy (brain dysfunction) Nausea Vomiting	Anemia Encephalopathy Foot drop/wrist drop (palsy) Nephropathy (kidney disease)

Chromium	Gastrointestinal hemorrhage (bleeding) Hemolysis (red blood cell destruction) Acute renal failure	Pulmonary fibrosis (lung scarring) Lung cancer
Arsenic	Nausea Vomiting Diarrhea Encephalopathy Multi-organ effects Arrhythmia Painful neuropathy	Diabetes Hypopigmentation/Hyperkeratosis Cancer

The toxic effects of arsenic, mercury and lead were known to the ancients but methodical studies of the overall toxicity of heavy metals appear to date from only 1868. In that year, Wanklyn and Chapman speculated on the adverse effects of the heavy metals "arsenic, lead, copper, zinc, iron and manganese" in drinking water. They noted an "absence of investigation" and were reduced to "the necessity of pleading for the collection of data." In 1884, Blake described an apparent connection between toxicity and the atomic weight of an element. The following sections provide historical thumbnails for the "classical" toxic heavy metals (arsenic, mercury and lead) and some more recent examples (chromium and cadmium).

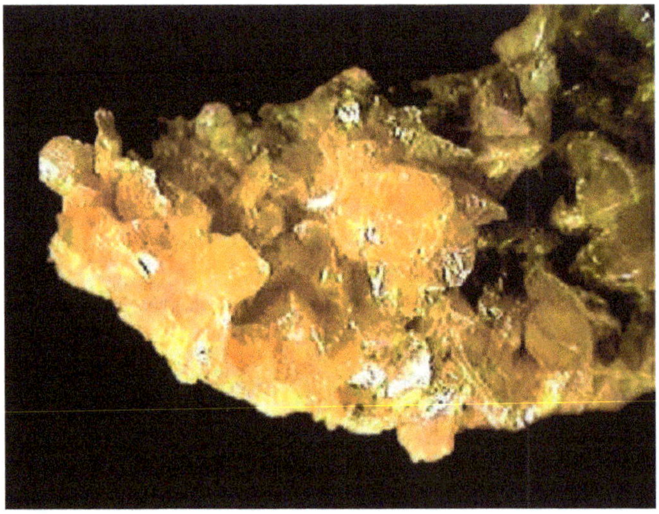

Orpiment, a toxic arsenic mineral used in the tanning industry to remove hair from hides.

Arsenic

Arsenic, as realgar (As_4S_4) and orpiment (As_2S_3), was known in ancient times. Strabo (64–50 BCE – c. AD 24?), a Greek geographer and historian, wrote that only slaves were employed in realgar and orpiment mines since they would inevitably die from the toxic effects of the fumes given off from the ores. Arsenic-contaminated beer poisoned over 6,000 people in the Manchester area of England in 1900, and is thought to have killed at least 70 victims. Clare Luce, American ambassador to Italy from 1953 to 1956, suffered from arsenic poisoning. Its source was traced to flaking arsenic-laden paint on the ceiling of her bedroom. She may also have eaten food contaminated by arsenic in flaking ceiling paint in the embassy dining room. Ground water contaminated by arsenic, as of 2014, "is still poisoning millions of people in Asia."

Mercury

Saint Isaac's Cathedral, in Saint Petersburg, Russia. The gold-mercury amalgam used to gild its dome caused numerous casualties among the workers involved.

The first emperor of unified China, Qin Shi Huang, it is reported, died of ingesting mercury pills that were intended to give him eternal life. The phrase "mad as a hatter" is likely a reference to mercury poisoning among milliners (so-called "mad hatter disease"), as mercury-based compounds were once used in the manufacture of felt hats in the 18th and 19th century. Historically, gold amalgam (an alloy with mercury) was widely used in gilding, leading to numerous casualties among the workers. It is estimated that during the construction of Saint Isaac's Cathedral alone, 60 workers died from the gilding of the main dome.Outbreaks of methylmercury poisoning occurred in several places in Japan during the 1950s due to industrial discharges of mercury into rivers and coastal waters. The best-known instances were in Minamata and Niigata. In Minamata alone, more than 600 people died due to what became known as Minamata disease. More than 21,000 people filed claims with the Japanese government, of which almost 3000 became certified as having the disease. In 22 documented cases, pregnant women who consumed contaminated fish showed mild or no symptoms but gave birth to infants with severe developmental disabilities. Since the industrial Revolution, mercury levels have tripled in many near-surface seawaters, especially around Iceland and Antarctica.

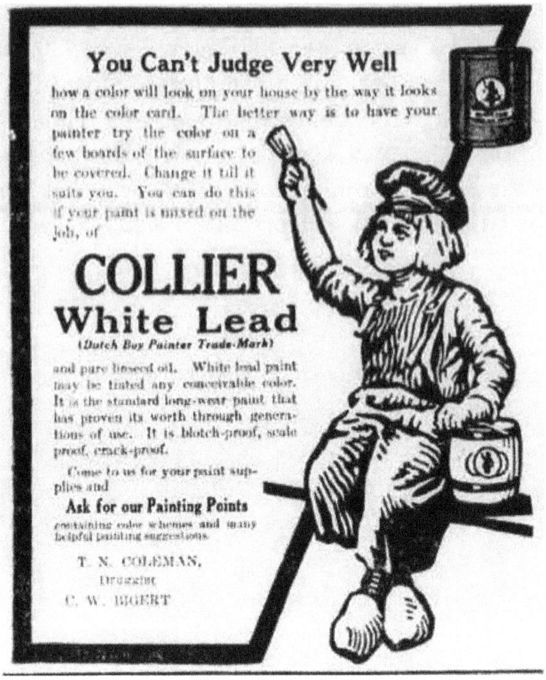

Dutch Boy white lead paint advertisement, 1912.

Lead

The adverse effects of lead were known to the ancients. In the 2nd century BC the Greek botanist Nicander described the colic and paralysis seen in lead-poisoned people. Dioscorides, a Greek physician who is thought to have lived in the 1st century CE, wrote that lead "makes the mind give way". Lead was used extensively in Roman aqueducts from about 500 BC to 300 AD. Julius Caesar's engineer, Vitruvius, reported, "water is much more wholesome from earthenware pipes than from lead pipes. For it seems to be made injurious by lead, because white lead is produced by it, and this is said to be harmful to the human body." During the Mongol period in China (1271–1368 AD), lead pollution due to silver smelting in the Yunnan region exceeded contamination levels from modern mining activities by nearly four times. In the 17th and 18th centuries, people in Devon were afflicted by a condition referred to as Devon colic; this was discovered to be due to the imbibing of lead-contaminated cider. In 2013, the World Health Organization estimated that lead poisoning resulted in 143,000 deaths, and "contributed to 600,000 new cases of children with intellectual disabilities", each year. In the U.S. city of Flint, Michigan, lead contamination in drinking water has been an issue since 2014. The source of the contamination has been attributed to "corrosion in the lead and iron pipes that distribute water to city residents". In 2015, drinking water lead levels in north-eastern Tasmania, Australia, were reported to reach over 50 times national drinking water guidelines. The source of the contamination was attributed to "a combination of dilapidated drinking water infrastructure, including lead jointed pipelines, end-of-life polyvinyl chloride pipes and household plumbing."

Chromium

Potassium chromate, a carcinogen, is used in the dyeing of fabrics, and as a tanning agent to produce leather.

Chromium(III) compounds and chromium metal are not considered a health hazard, while the toxicity and carcinogenic properties of chromium(VI) have been known since at least the late 19th century. In 1890, Newman described the elevated cancer risk of workers in a chromate dye company. Chromate-induced dermatitis was reported in aircraft workers during World War II. In 1963, an outbreak of dermatitis, ranging from erythema to exudative eczema, occurred amongst 60 automobile factory workers in England. The workers had been wet-sanding chromate-based primer paint that had been applied to car bodies.In Australia, chromium was released from the Newcastle Orica explosives plant on August 8, 2011. Up to 20 workers at the plant were exposed as were 70 nearby homes in Stockton. The town was only notified three days after the release and the accident sparked a major public controversy, with Orica criticised for playing down the extent and possible risks of the leak, and the state Government attacked for their slow response to the incident.

Cadmium

Cadmium exposure is a phenomenon of the early 20th century, and onwards. In Japan in 1910, the Mitsui Mining and Smelting Company began discharging cadmium into the Jinzugawa river, as a byproduct of mining operations. Residents in the surrounding area subsequently consumed rice grown in cadmium-contaminated irrigation water. They experienced softening of the bones and kidney failure. The origin of these symptoms was not clear; possibilities raised at the time included "a regional or bacterial disease or lead poisoning." In 1955, cadmium was identified as the likely cause and in 1961 the source was directly linked to mining operations in the area. In February 2010, cadmium was found in Walmart exclusive Miley Cyrus jewelry. Wal-Mart continued to sell the jewelry until May, when covert testing organised by Associated Press confirmed the original results. In June 2010 cadmium was detected in the paint used on promotional drinking glasses for the movie Shrek Forever After, sold by McDonald's Restaurants, triggering a recall of 12 million glasses.

Remediation

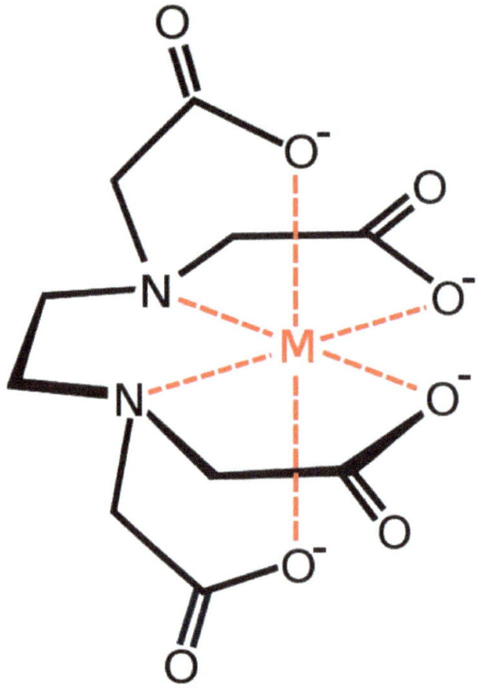

A metal EDTA anion. Pb displaces Ca in Na_2[CaEDTA] to give Na_2[PbEDTA], which is passed out of the body in urine.

In humans, heavy metal poisoning is generally treated by the administration of chelating agents. These are chemical compounds, such as CaNa2 EDTA (calcium disodium ethylenediaminetetraacetate) that convert heavy metals to chemically inert forms that can be excreted without further interaction with the body. Chelates are not without side effects and can also remove beneficial metals from the body. Vitamin and mineral supplements are sometimes co-administered for this reason.

Soils contaminated by heavy metals can be remediated by one or more of the following technologies: isolation; immobilization; toxicity reduction; physical separation; or extraction. *Isolation* involves the use of caps, membranes or below-ground barriers in an attempt to quarantine the contaminated soil. *Immobilization* aims to alter the properties of the soil so as to hinder the mobility of the heavy contaminants. *Toxicity reduction* attempts to oxidise or reduce the toxic heavy metal ions, via chemical or biological means into less toxic or mobile forms. *Physical separation* involves the removal of the contaminated soil and the separation of the metal contaminants by mechanical means. *Extraction* is an on or off-site process that uses chemicals, high-temperature volatization, or electrolysis to extract contaminants from soils. The process or processes used will vary according to contaminant and the characteristics of the site.

Benefits

Some elements otherwise regarded as toxic heavy metals are essential, in small quantities, for human health. These elements include vanadium, manganese, iron, cobalt, copper, zinc, selenium, strontium and molybdenum. A deficiency of these essential metals may increase susceptibility to heavy metal poisoning.

REFERENCES

1. "Frost – Definition of frost by Merriam-Webster". merriam-webster.com.
2. John E. Oliver (1 January 2005). The Encyclopedia of World Climatology. Springer Science & Business Media. pp. 382–. ISBN 978-1-4020-3264-6.
3. "What causes frost?". Retrieved 2007-12-05.
4. "Hoarfrost – Definition of hoarfrost by Merriam-Webster". merriam-webster.com.
5. Louisiana State University Baton Rouge Robert V Rohli; Robert V. Rohli; Anthony J. Vega (13 December 2013). Climatology. Jones & Bartlett Publishers. pp. 381–. ISBN 978-1-284-05427-9.
6. "Weather Facts: Frost hollow – Weather UK – weatheronline.co.uk". weatheronline.co.uk.
7. David McClung; Peter A. Schaerer (2006). The Avalanche Handbook. The Mountaineers Books. pp. 72–. ISBN 978-0-89886-809-8.
8. West, Bruce; Mauro Bologna (2003). Physics of Fractal Operators. Paolo Grigolini. Springer. p. 46. ISBN 0-387-95554-2.
9. "Rime – Definition of rime by Merriam-Webster". merriam-webster.com.
10. "Frost Tolerance of Vegetables". Botanical Interests. Retrieved Nov 12, 2013.
11. "Fall vegetables vs. Summer vegetables".
12. Klosinska, Urszula; et al. (27 February 2014), "Low temperature seed germination of cucumber: genetic basis of the tolerance trait", Journal of Horticultural Research, de Gruyter, 21 (2), doi:10.2478/johr-2013-0031, retrieved 2014-02-28
13. Maki LR, Galyan EL, Chang-Chien MM, Caldwell DR (1974). "Ice Nucleation Induced by Pseudomonas syringae". Applied Microbiology. 28 (3): 456–459. PMC 186742 . PMID 4371331.
14. Lindow, Stephen E.; Deane C. Arny; Christen D. Upper (October 1982). "Bacterial

Ice Nucleation: A Factor in Frost Injury to Plants". Plant Physiology. 70 (4): 1084–1089. doi:10.1104/pp.70.4.1084. PMC 1065830 . PMID 16662618.

15. Selective Inverted Sink Rolex Awards site (won award in Technology and Innovation category) 1998.
16. Helicopters Fight Frost – Vector, Civil Aviation Authority of New Zealand, September / October 2008, Page 8-9
17. Turbines and turbulence, Nature (journal), 468, 1001, 23 December 2010, DOI:10.1038/4681001a, published online 22 December 2010.
18. Somnath Baidya Roy and Justin J. Traiteur. Impacts of wind farms on surface air temperatures, Proceedings of the National Academy of Sciences, Vol. 107, No. 42, October 19, 2010, p. 17,899.
19. Wind farms impacting weather, Science Daily.
20. "A practical method of frost protection". Retrieved 31 October 2011.
21. Selders, Arthur W. "Frost protection with sprinkler irrigation"(PDF). West Virginia university. Retrieved 31 October 2011.
22. Long S.P., Farage P.K., Garcia R.L., (1996) Measurement of leaf and canopy photosynthetic C02 exchange in the field, Journal of Experimental Botany, Vol. 47, No. 304, pp. 1629-1642
23. Baker N.R. (2008) Chlorophyll Fluorescence: A Probe of Photosynthesis In Vivo, Annu. Rev. Plant Biol.2008. 59:89–113
24. Long S.P., and Bernacchi C.J. (2003) Gas exchange measurements, what can they tell us about the underlying limitations to photosynthesis? Procedures and sources of error Journal of Experimental Botany, Page 1 of 9
25. Edwards GE and Baker NR (1993)Can CO2 assimilation in maize leaves be predicted accurately from chlorophyll fluorescence analysis? Photosynth Res 37: 89–102
26. Baker N. R., Oxborough K., (2004) Chlorophyll fluorescence as a probe of photosynthetic productivity. From Chapter 3, "Chlorophyll a Fluorescence a Signature of Photosynthesis", edited by George Papaqeorgiou and Govindjee, published by Springer 2004, PO Box 17, 3300 AA Dordrecht, The Netherlands, pages 66-79
27. Fryer M. J., Andrews J.R., Oxborough K., Blowers D. A., Baker N.E. (1998) Relationship between CO2 Assimilation, Photosynthetic Electron Transport, and Active O2 Metabolism in Leaves of Maize in the Field during Periods of Low Temperature. Plant Physiol. (1998) 116:571–580
28. Rosyara, U. R., S. Subdedi, R. C. Sharma and E. Duveiller. 2010. Photochemical Efficiency and SPAD Value as Indirect Selection Criteria for Combined Selection of Spot Blotch and Terminal Heat Stress in Wheat. Journal of Phytopathology Volume 158, Issue 11-12, pages 813–821
29. Maxwell K., Johnson G. N, (2000) Chlorophyll fluorescence – a practical guide. Journal of Experimental Botany Vol. 51, No. 345, pp. 659-668- April 2000

30. Baker N.R, Rosenquist E. (2004) Applications of chlorophyll fluorescence can improve crop production strategies: an examination of future possibilities, Journal of Experimental Botany, 55(403):1607-1621

31. Kalaji M.H., Govindjee, Bosa K., Kościelniak J., Żuk-Gołaszewska K., 2011. Effects of salt stress on Photosystem II efficiency and CO2 assimilation of two Syrian barley landraces. Environ. Exper. Bot. 73: 64–72.

32. Kalaji M.H., Bosa K., Kościelniak J., Hossain Z. 2011. Chlorophyll a fluorescence - A useful tool for the early detection of temperature stress in spring barley (Hordeum vulgare L.). OMICS A Journal of Integrative Biology 15 (12): 925–934.

33. Bussotti F., Desotgiu R., Pollastrini M., Kalaji M.H., Łoboda T., Bosa K., 2012. Misurare la vitalità delle piante per mezzo della fluorescenza della clorofilla. Università di Firenze. I edizione italiana.

34. Kitajima M, Butler WL (1975) Quenching of chlorophyll fluorescence and primary photochemistry in chloroplasts by dibromothymoquinone. Biochim Biophys Acta 376:105-115

35. Genty B., Briantais J. M. & Baker N. R. (1989) The relationship between the quantum yield of photosynthetic electron transport and quenching of chlorophyll fluorescence, Biochimica et Biophysica Acta 990, 87-92

36. Genty B., Harbinson J., Baker N.R. (1990) Relative quantum efficiencies of the two photosystems of leaves in photo respiratory and non-photo respiratory conditions. Plant Physiol Biochem 28: 1-10

37. Laisk A and Loreto F (1996) Determining photosynthetic parameters from leaf CO2 exchange and chlorophyll fluorescence. Ribulose-1,5-bisphosphate carboxylase / oxygenase specificity factor, dark respiration in the light, excitation distribution between photosystems, alternative electron transport rate, and mesophyll diffusion resistance. Plant Physiol110: 903–912.

38. Eichelman H, Oja V., Rasulov B., Padu E., Bichele I., Pettai H., Niinemets O., Laisk A. (2004) Development of Leaf Photosynthetic Parameters in Betual pendula Roth Leaves: Correlation with Photosystem I Density, Plant Biology 6 (2004): 307-318

39. Schreiber U, (2004) Pulse-Amplitude-Modulation (PAM) Fluorometry and Saturation Pulse Method: An Overview From Chapter 11, "Chlorophyll a Fluorescence a Signature of Photosynthesis", edited by George Papaqeorgiou and Govindjee, published by Springer 2004,PO Box 17, 3300 AA Dordrecht, The Netherlands, page 279–319

40. Flexas (2000)– "Steady-State and Maximum Chlorophyll Fluorescence Responses to Water StressIn Grape Vine Leaves: A New Remote Sensing System", J. Flexas, MJ Briantais, Z Cerovic, H Medrano, I Moya, Remote Sensing Environment 73:283-270

41. "Quenching application note". Optisci.com. Retrieved 2012-03-08.

42. Kramer D. M., Johnson G., Kiirats O., Edwards G. (2004) New fluorescence parameters for determination of QA redox state and excitation energy fluxes. Photosynthesis Research 79: 209-218

43. Klughammer C., and Schreiber U. (2008) PAM Application notes 2008 1:27 -35

44. Hendrickson L., Furbank R., & Chow (2004) A simple alternative approach to assessing the fate of absorbed Light energy using chlorophyll fluorescence. Photosynthesis Research 82: 73-81

45. Strasser R.J. Tsimilli-Michael M., and Srivastava A. (2004) - Analysis of Chlorophyll a Fluorescence Transient. From Chapter 12, "Chlorophyll a Fluorescence a Signature of Photosynthesis", edited by George Papaqeorgiou and Govindjee, published by Springer 2004, PO Box 17, 3300 AA Dordrecht, The Netherlands, page 340

46. . Vredenberg, W.J., (2004) System analysis of photoelectrochemical control of chlorophyll fluorescence in terms of trapping models of Photosystem II: a challenging view. In: Papageorgiou, G.C., Govindjee(Eds.), Chlorophyll a Fluorescence: A Signature of Photosynthesis, Advances in Photosynthesis and Respiration, vol. 19. Springer, Dordrecht, pp. 133–172.

47. . Vredenberg, W.J., Prasil, O., (2009) Modeling of chlorophyll a fluorescence kinetics in plant cells: derivation of a descriptive algorithm. In: Laisk, A., Nedbal, L., Govindjee(Eds.), Photosynthesis in Silico: Understanding Complexity from Molecules to Ecosystems. Springer, Dordrecht (the Netherlands), pp. 125–149.

48. Vredenburg, W.J (2011) Kinetic analyses and mathematical modeling of primary photochemical and photoelectrochemical processes in plant photosystems, BioSystems 103 (2011) 138–151

49. Haldimann P, & Feller U. (2004) Inhibition of photosynthesis by high temperature in oak (Quercus pubescens L.) leaves grown under natural conditions closely correlates with a reversible heat dependent reduction of the activation state of ribulose-1,5-bisphosphatecarboxylase/oxygenase.

50. Rosenqvist E., van Kooten O., (2006) Chlorophyll Fluorescence: A General Description and Nomenclature. From Chapter 2 "Practical Applications of Chlorophyll Fluorescence in Plant Biology". by Jennifer R. DeEll (Editor), Peter M.A. Toivonen (Editor) Kluwer

51. Cavender-Bares J. & Fakhri A. Bazzaz (2004)– "From Leaves to Ecosystem: Using Chlorophyll Fluorescence to Assess Photosynthesis and Plant Function in Ecological Studies". Jeannine Cavender Bares, Fakhri A. Bazzaz, From Chapter 29, "Chlorophyll a Fluorescence a Signature of Photosynthesis", edited by George Papaqeorgiou and Govindjee, published by Springer 2004, PO Box 17, 3300 AA Dordrecht, The Netherlands, page 746-747

52. da Silva J. A. & Arrabaca M.C. (2008). Photosynthesis in the water-stressed C4 grass Setaria sphacelata is mainly limited by stomata with both rapidly and slowly imposed water deficits. Physiologia Plantarum Volume 121 Issue 3, Pages 409 – 420 2008

53. Burke J. (2007) Evaluation of Source Leaf Responses to Water-Deficit Stresses in Cotton Using a Novel Stress Bioassay, Plant Physiology, Jan. 2007, Vol 143, pp108-121

54. Burke J., Franks C.D. Burow G., Xin Z. (2010) Selection system for the Stay-Green DroughtTolerance Trait in Sorghum Germplasm, Agronomy Journal 102:1118-1122 May 2010
55. Knighton N., Bugbee B., (2004)– A comparison of Opti-Sciences CCM-200 chlorophyll meter and the Minolta SPAD 502 chlorophyll meter, Crop Physiology Laboratory - Utah State University
56. Richardson A. D., Duigan S.P., Berlyn G.P., (2002) An evaluation of noninvasive methods to estimate foliar chlorophyll content New Phytologist (2002)153 : 185–194
57. Shapiro C., Schepers J., Francis D., Shanahan J., (2006) Using a Chlorophyll Meter to improve N Management. NebGuide # 1621 University of Nebraska – Lincoln Extension
58. Francis D.D., Piekielek D.D. Francis and W.P. Piekielek. SSMG-12. Introduction. Determining when to apply additional N ... Assessing Crop Nitrogen Needs with. Chlorophyll Meters
59. Cheng L., Fuchigami L., Breen P., (2001) "The relationship between photosystem II efficiency and quantum yield for CO2 assimilation is not affected by nitrogen content in apple leaves."Journal of Experimental Botany, 52(362):1865-1872
60. Perry, Leonard. "pH for the Garden". Retrieved 11 December 2012.
61. Soil Survey Division Staff. "Soil survey manual.1993. Chapter 3, selected chemical properties.". Soil Conservation Service. U.S. Department of Agriculture Handbook 18. Retrieved 2011-03-12.
62. Sparks, Donald; Environmental Soil Chemistry. 2003, Academic Press, London, UK
63. http://edis.ifas.ufl.edu/ch086
64. Finck, Arnold (1976). Pflanzenernährung in Stichworten. Kiel: Hirt. p. 80. ISBN 3-554-80197-6.
65. Brady, N. and Weil, R. The Nature and Properties of Soils. 13th ed. 2002
66. Hansson et al (2011) Differences in soil properties in adjacent stands of Scots pine, Norway spruce and silver birch in SW Sweden. Forest Ecology and Management 262 522–530
67. Rout, GR; Samantaray, S; Das, P (2001). "Aluminium toxicity in plants: a review" (PDF). Agronomie. 21 (1): 4–5. doi:10.1051/agro:2001105. Retrieved 11 June 2014.
68. http://www.extension.org/pages/9875/soil-ph-and-nutrient-availability
69. Soilinfo PSU
70. USDA Myakka data
71. USDA Berkshire data
72. USDA Cataloochee data
73. Buol, S. W., R. J. Southard, R.C. Graham and P.A. McDaniel. Soil Genesis and Classification. (5th) Edition, Ia. State Press p. 494. 2002

74. http://www2.hawaii.edu/~nvhue/acid.html
75. Brady, N. and Weil, R. The Nature and Properites of Soils. 13th ed. 2002
76. Aggrawal, A. Textbook of Forensic Medicine and Toxicology. New Dehli: Avichal Publishing Company. ISBN 978-81-7739-419-1.}
77. Balasubramanian, R; He, J; Wang, LK (2009). "Control, Management, and Treatment of Metal Emissions from Motor Vehicles". In Shammas, LK; Wang, JP; Chen, Y; et al. Heavy Metals in the Environment. CRC Press. pp. 475–490. ISBN 1420073168.
78. Baldwin DR, Marshall WJ (1999). "Heavy metal poisoning and its laboratory investigation". Ann Clin Biochem. 36 (3): 267–300. doi:10.1177/000456329903600301. PMID 10376071.
79. Bánfalvi, G (2011). "Heavy Metals, Trace Elements and their Cellular Effects". In Bánfalvi, G. Cellular Effects of Heavy Metals. Springer. pp. 3–28. ISBN 9789400704275.
80. Barceloux DG (1999). "Chromium". J Toxicol Clin Toxicol. 37 (2): 173–94. doi:10.1081/CLT-100102418. PMID 10382554.
81. Blake J (1884). "On the Connection Between Physiological Action and Chemical Constitution". The Journal of Physiology. 5 (1): 36–44. doi:10.1113/jphysiol.1884.sp000148. PMC 1484879 . PMID 16991361.
82. Brathwaite RL, Rabone SD (1985). "Heavy Metal Sulphide Deposits and Geochemical Surveys for Heavy Metals in New Zealand". Journal of the Royal Society of New Zealand. 15 (4): 363–370. doi:10.1080/03036758.1985.10421713.
83. Cameron CA (1871). "Half-yearly Report on Public Health". Dublin Quarterly Journal of Medical Science. 52 (2): 475–498. doi:10.1007/BF02944536.
84. "Chelation Therapy". American Cancer Society. 2008. Retrieved 2014-04-28.
85. "Chelation: Therapy or "Therapy"?". National Capital Poison Center. 2010.
86. Chowdhury BA, Chandra RK (1987). "Biological and health implications of toxic heavy metal and essential trace element interactions". Prog Food Nutr Sci. 11 (1): 55–113. PMID 3303135.
87. Cole M.; Lindeque P.; Halsband C.; Galloway T.S. (2011). "Microplastics as contaminants in the marine environment: A review". Marine Pollution Bulletin. 62 (12): 2588–2597. doi:10.1016/j.marpolbul.2011.09.025.
88. Csuros M (1997). Environmental Sampling and Analysis Lab Manual. Lewis. ISBN 1566701783.
89. Davidson PW, Myers GJ, Weiss B (2004). "Mercury exposure and child development outcomes". Pediatrics. 113: 1023–9. doi:10.1542/peds.113.4.S1.1023. PMID 15060195.
90. Dewan S (26 December 2008). "Tennessee Ash Flood Larger Than Initial Estimate". New York Times.
91. Dewan S (1 January 2009). "Metal Levels Found High in Tributary After Spill". New York Times.

92. Di Maio, VJM (2001). Forensic Pathology (2nd ed.). CRC Press. ISBN 084930072X.
93. Dueck, D (2000). Strabo of Amasia: A Greek Man of Letters in Augustan Rome. Routledge. ISBN 0415216729.
94. Duffus JH (2002). "Heavy Metals"—A Meaningless Term?". Pure and Applied Chemistry. 74 (5): 793–807. doi:10.1351/pac200274050793.
95. Dyer P (2009). "The 1900 Arsenic Poisoning Epidemic". Brewery History (130): 65–85.
96. Emsley, J (2011). Nature's Building Blocks. Oxford University Press. ISBN 9780199605637.
97. Evanko, CA; Dzombak, DA (1997). "Remediation of Metals-Contaminated Soils and Groundwater". Technology Evaluation Report. Pittsburgh PA: Ground-water Remediation Technologies Center. CiteSeerX 10.1.1.401.3768 . TE 97-0-1.
98. Finch, LE; Hillyer, MM; Leopold, MC (2015). "Quantitative Analysis of Heavy Metals in Children's Toys and Jewelry: A Multi-Instrument Multitechnique Exercise in Analytical Chemistry and Public Health". Journal of Chemical Education. 92 (5): pp. 849–854. doi:10.1021/ed500647w.
99. Gilbert SG, Weiss B (2006). "A Rationale for Lowering the Blood Lead Action Level from 10 to 2 μg/dL". Neurotoxicology. 27 (5): 693–701. doi:10.1016/j.neuro.2006.06.008. PMC 2212280 . PMID 16889836.
100. Haines, AT; Nieboer, E (1988). "Chromium hypersensitivity". In Nriagu, JO; Nieboer, E. Chromium in the Natural and Human Environments. Wiley. pp. 497–532. ISBN 0471856436.
101. Harvey PJ, Handley HK, Taylor MP (April 2015). "Identification of the sources of metal (lead) contamination in drinking waters in north-eastern Tasmania using lead isotopic compositions". Environmental Science and Pollution Research. 22: 12276–12288. doi:10.1007/s11356-015-4349-2.
102. Hawkes SJ (1997). "What is a "Heavy Metal"?". Journal of Chemical Education. 74 (11): 1374. doi:10.1021/ed074p1374.
103. Hillman AL, Abbot MB, Yu JQ, Bain DJ, Chiou-Peng TH (2015). "Environmental Legacy of Copper Metallurgy and Mongol Silver Smelting Recorded in Yunnan Lake Sediments". Environmental Science & Technology. 49 (6): 3349–3357. doi:10.1021/es504934r.
104. "Heavy Metal Poisoning". National Organization for Rare Disorders. 2015. Retrieved 11 February 2016.
105. Heavy Metal Toxicity at eMedicine
106. Houlton S (2014). "Boom!". Chemistry World. 11 (12): 48–51.
107. Howell N, Lavers J, Paterson D, Garrett R, Banati R (2012). "Trace metal distribution in feathers from migratory, pelagic birds". Australian Nuclear Science and Technology Organisation. Retrieved 2014-05-03.
108. Jones J (11 August 2011). "Stockton Residents Fume Over Fallout From Orica". Newcastle Herald. Retrieved 2014-05-16.

109. Landis, WG; Sofield, RM; Yu, M-H (2000). Introduction to Environmental Toxicology: Molecular Substructures to Ecological Landscapes. 4th: CRC Press. ISBN 9781439804100.

110. Lovei, M (1998). Phasing Out Lead from Gasoline: Worldwide Experience and Policy Implications. World Bank Technical Paper. 397. The World Bank. ISBN 082134157X. ISSN 0253-7494.

111. Mulvihill G, Pritchard J (4 June 2010). "McDonald's Recall: ‹Shrek' Glasses Contain Toxic Metal Cadmium". Huffington Post.

112. National Research Council (U.S.). Committee on Biologic Effects of Atmospheric Pollutants (1974). Chromium. National Academies. NAP:13852.

113. Needleman H (2004). "Lead poisoning". Annu Rev Med. 55: 209–22. doi:10.1146/annurev.med.55.091902.103653. PMID 14746518.

114. Newman D (1890). "A Case of Adeno-carcinoma of the Left Inferior Turbinated Body, and Perforation of the Nasal Septum, in the Person of a Worker in Chrome Pigments". The Glasgow Medical Journal. 33: 469–470.

115. Nielen, MWF; Marvin, HJP (2008). "Challenges in Chemical Food Contaminants and Residue Analysis". In Picó, Y. Food Contaminants and Residue Analysis. Elsevier. pp. 1–28. ISBN 0080931928.

116. Notman N (2014). "Digging Deep for Safer Water". Chemistry World. 11 (4): 54–57.

117. O'Brien N, Aston H (13 November 2011). "The untold story of Orica's chemical leaks". Sydney Morning Herald.

118. Pearce JM (2007). "Burton's Line in Lead Poisoning". Eur Neurol. 57 (2): 118–9. doi:10.1159/000098100. PMID 17179719.

119. Perry, J; Vanderklein, EL (1996). Water Quality: Management of a Natural Resource. Blackwell Science. ISBN 0865424691.

120. Pezzarossa, B; Gorini, F; Petruzelli, G (2011). "Heavy Metal and Selenium Distribution and Bioavailability in Contaminated Sites: A Tool for Phytoremediation". In Selim, HM. Dynamics and Bioavailabiliy of Heavy Metals in the Rootzone. CRC Press. pp. 93–128. ISBN 9781439826225.

121. Poovey B (15 September 2001). "Trial Starts on Damage Lawsuits in TVA Ash Spill". Bloomberg Businessweek.

122. Prioreschi, P (1998). Roman Medicine. A History of Medicine. III. Horatius Press. ISBN 1888456035.

123. Pritchard J (19 May 2010). "Wal-Mart Pulls Miley Cyrus Jewelry After Cadmium Tests". USA Today.

124. Qu, C; Ma, Z; Yang, J; Lie, Y; Bi, J; Huang, L (2014). "Human Exposure Pathways of Heavy Metal in a Lead-Zinc Mining Area". In Asrari, E. Heavy Metal Contamination of Water and Soil: Analysis, assessment, and remediation strategies. Apple Academic Press. pp. 129–156. ISBN 9781771880046.

125. Radojevic, M; Bashkin, VN (1999). Practical Environmental Analysis. Royal Society of Chemistry. ISBN 0854045945.

126. Rand, GM; Wells, PG; McCarty, LS (1995). "Introduction to aquatic toxicology". In Rand, GM. Fundamentals Of Aquatic Toxicology: Effects, Environmental Fate And Risk Assessment (2nd ed.). Taylor & Francis. pp. 3–70. ISBN 1560320907.

127. Rogers, MJ (2000). "Text and Illustrations. Dioscorides and the Illuminated Herbal in the Arab Tradition". In Contadini, A. Arab Painting: Text and Image in Illustrated Arabic Manuscripts. Leiden: Koninklijke Brill NV. pp. 41–48 (41). ISBN 9789004186309.

128. "Rogue mercury". New Scientist. 223 (2981). 2014.

129. Sengupta, AK (2002). "Principles of Heavy Metals Separation". In Sengupta, AK. Environmental Separation of Heavy Metals: Engineering Processes. Lewis. ISBN 1566768845.

130. Srivastava, S; Goyal, P (2010). Novel Biomaterials: Decontamination of Toxic Metals from Wastewater. Springer-Verlag. ISBN 978-3-642-11329-1.

131. "Stop Lead Poisoning in Children". World Health Organization. 2013.

132. "Ten Chemicals of Major Public Health Concern". World Health Organisation. 2015.

133. Torrice, M. "How Lead Ended Up In Flint's Tap Water". Chemical & Engineering News. 94 (7): 26–29.

134. Tovey J (17 December 2011). "Patches of Carcinogen Seen After Orica Leak". The Sydney Morning Herald.

135. The United States Pharmacopeia (21st ed.). The United States Pharmacopeial Convention. 1985. ISBN 0-913595-04-7.

136. Vallero, DA; Letcher, TM (2013). Unravelling environmental disasters. Elsevier. ISBN 9780123970268.

137. Waldron HA (1983). "Did the Mad Hatter have Mercury Poisoning?". Br Med J (Clin Res Ed). 287 (6409): 1961. doi:10.1136/bmj.287.6409.1961. PMC 1550196 . PMID 6418283.

138. Wanklyn, JA; Chapman, ET (1868). Water-analysis: A Practical Treatise on the Examination of Potable Water. Trüber & Company.

139. Whorton, JG (2011). The Arsenic Century. Oxford University Press. ISBN 9780199605996.

140. Worsztynowicz, A; Mill, W (1995). "Potential Ecological Risk due to Acidification of Heavy Industrialized Areas — The Upper Silesia Case". In Erisman, JW; Hey, GJ. Acid Rain Research: Do We Have Enough Answers?,. Elsevier. pp. 353–66. ISBN 0444820388.

141. Wright, DA; Welbourn, P (2002). Environmental Toxicology. Cambridge University Press. ISBN 0521581516.

142. Zhao HL, Zhu X, Sui Y (2006). "The short-lived Chinese emperors". J Am Geriatr Soc. 54 (8): 1295–6. doi:10.1111/j.1532-5415.2006.00821.x. PMID 16914004.

143. Ahrland S., Liljenzin J. O. & Rydberg J. 1973, "Solution chemistry," in J. C. Bailar & A. F. Trotman-Dickenson (eds), Comprehensive Inorganic Chemistry, vol. 5, The Actinides, Pergamon Press, Oxford.

144. Albutt M. & Dell R. 1963, The nitrites and sulphides of uranium, thorium and plutonium: A review of present knowledge, UK Atomic Energy Authority Research Group, Harwell, Berkshire.

145. Alves A. K., Berutti, F. A. & Sánche, F. A. L. 2012, "Nanomaterials and catalysis", in C. P. Bergmann & M. J. de Andrade (ads), Nanonstructured Materials for Engineering Applications, Springer-Verlag, Berlin, ISBN 978-3-642-19130-5.

146. Amasawa E., Yi Teah H., Yu Ting Khew, J., Ikeda I. & Onuki M. 2016, "Drawing Lessons from the Minamata Incident for the General Public: Exercise on Resilience, Minamata Unit AY2014", in M. Esteban, T. Akiyama, C. Chen, I. Ikea, T. Mino (eds), Sustainability Science: Field Methods and Exercises, Springer International, Switzerland, pp. 93–116, doi:10.1007/978-3-319-32930-7_5 ISBN 978-3-319-32929-1.

147. Ariel E., Barta J. & Brandon D. 1973, "Preparation and properties of heavy metals", Powder Metallurgy International, vol. 5, no. 3, pp. 126–129.

148. Atlas R. M. 1986, Basic and Practical Microbiology, Macmillan Publishing Company, New York, ISBN 978-0-02-304350-5.

149. Australian Government 2016, National Pollutant Inventory, Department of the Environment and Energy, accessed 16 August 2016.

150. Baird C. & Cann M. 2012, Environmental Chemistry, 5th ed., W. H. Freeman and Company, New York, ISBN 978-1-4292-7704-4.

151. Baldwin D. R. & Marshall W. J. 1999, "Heavy metal poisoning and its laboratory investigation", Annals of Clinical Biochemistry, vol. 36, no. 3, pp. 267–300, doi:10.1177/000456329903600301.

152. Ball J. L., Moore A. D. & Turner S. 2008, Ball and Moore's Essential Physics for Radiographers, 4th ed., Blackwell Publishing, Chichester, ISBN 978-1-4051-6101-5.

153. Bánfalvi G. 2011, "Heavy metals, trace elements and their cellular effects", in G. Bánfalvi (ed.), Cellular Effects of Heavy Metals, Springer, Dordrecht, pp. 3–28, ISBN 978-94-007-0427-5.

154. Baranoff E. 2015, "First-row transition metal complexes for the conversion of light into electricity and electricity into light", in W-Y Wong (ed.), Organometallics and Related Molecules for Energy Conversion, Springer, Heidelberg, pp. 61–90, ISBN 978-3-662-46053-5.

155. Berea E., Rodriguez-lbelo M. & Navarro J. A. R. 2016, "Platinum Group Metal—Organic frameworks" in S. Kaskel (ed.), The Chemistry of Metal-Organic Frameworks: Synthesis, Characterisation, and Applications, vol. 2, Wiley-VCH Weinheim, pp. 203–230, ISBN 978-3-527-33874-0.

156. Berger A. J. & Bruning N. 1979, Lady Luck's Companion: How to Play ... How to Enjoy ... How to Bet ... How to Win, Harper & Row, New York, ISBN 978-0-06-014696-2.

157. Berry L. G. & Mason B. 1959, Mineralogy: Concepts, Descriptions, Determinations, W. H. Freeman and Company, San Francisco.

158. Biddle H. C. & Bush G. L 1949, Chemistry Today, Rand McNally, Chicago.

159. Bonchev D. & Kamenska V. 1981, "Predicting the properties of the 113–120 transactinide elements", The Journal of Physical Chemistry, vo. 85, no. 9, pp. 1177–1186, doi:10.1021/j150609a021.

160. Bonetti A., Leone R., Muggia F. & Howell S. B. (eds) 2009, Platinum and Other Heavy Metal Compounds in Cancer Chemotherapy: Molecular Mechanisms and Clinical Applications, Humana Press, New York, ISBN 978-1-60327-458-6.

161. Booth H. S. 1957, Inorganic Syntheses, vol. 5, McGraw-Hill, New York.

162. Bradl H. E. 2005, "Sources and origins of heavy metals", in Bradl H. E. (ed.), Heavy Metals in the Environment: Origin, Interaction and Remediation, Elsevier, Amsterdam, ISBN 978-0-12-088381-3.

163. Brady J. E. & Holum J. R. 1995, Chemistry: The Study of Matter and its Changes, 2nd ed., John Wiley & Sons, New York, ISBN 978-0-471-10042-3.

164. Brephohl E. & McCreight T. (ed) 2001, The Theory and Practice of Goldsmithing, C. Lewton-Brain trans., Brynmorgen Press, Portland, Maine, ISBN 978-0-9615984-9-5.

165. Brown I. 1987, "Astatine: Its organonuclear chemistry and biomedical applications," in H. J. Emeléus & A. G. Sharpe (eds), Advances in Inorganic Chemistry, vol. 31, Academic Press, Orlando, pp. 43–88, ISBN 978-0-12-023631-2.

166. Bryson R. M. & Hammond C. 2005, "Generic methodologies for nanotechnology: Characterisation"', in R. Kelsall, I. W. Hamley & M. Geoghegan, Nanoscale Science and Technology, John Wiley & Sons, Chichester, pp. 56–129, ISBN 978-0-470-85086-2.

167. Burkett B. 2010, Sport Mechanics for Coaches, 3rd ed., Human Kinetics, Champaign, Illinois, ISBN 978-0-7360-8359-1.

168. Casey C. 1993, "Restructuring work: New work and new workers in post-industrial production", in R. P. Coulter & I. F. Goodson (eds), Rethinking Vocationalism: Whose Work/life is it?, Our Schools/Our Selves Education Foundation, Toronto, ISBN 978-0-921908-15-9.

169. Chakhmouradian A.R., Smith M. P. & Kynicky J. 2015, "From "strategic" tungsten to "green" neodymium: A century of critical metals at a glance", Ore Geology Reviews, vol. 64, January, pp. 455–458, doi:10.1016/j.oregeorev.2014.06.008.

170. Chambers E. 1743, "Metal", in Cyclopedia: Or an Universal Dictionary of Arts and Sciences (etc.), vol. 2, D. Midwinter, London.

171. Chandler D. E. & Roberson R. W. 2009, Bioimaging: Current Concepts in Light & Electron Microscopy, Jones & Bartlett Publishers, Boston, ISBN 978-0-7637-3874-7.

172. Chawla N. & Chawla K. K. 2013, Metal matrix composites, 2nd ed., Springer Science+Business Media, New York, ISBN 978-1-4614-9547-5.

173. Chen J. & Huang K. 2006, "A new technique for extraction of platinum group

metals by pressure cyanidation", Hydrometallurgy, vol. 82, nos. 3–4, pp. 164–171, doi:10.1016/j.hydromet.2006.03.041.

174. Choptuik M. W., Lehner L. & Pretorias F. 2015, "Probing strong-field gravity through numerical simulation", in A. Ashtekar, B. K. Berger, J. Isenberg & M. MacCallum (eds), General Relativity and Gravitation: A Centennial Perspective, Cambridge University Press, Cambridge, ISBN 978-1-107-03731-1.

175. Clegg B 2014, "Osmium tetroxide", Chemistry World, accessed 2 September 2016.

176. Close F. 2015, Nuclear Physics: A Very Short Introduction, Oxford University Press, Oxford, ISBN 978-0-19-871863-5.

177. Clugston M & Flemming R 2000, Advanced Chemistry, Oxford University, Oxford, ISBN 978-0-19-914633-8.

178. Cole M., Lindeque P., Halsband C. & Galloway T. S. 2011, "Microplastics as contaminants in the marine environment: A review", Marine Pollution Bulletin, vol. 62, no. 12, pp. 2588–2597, doi:10.1016/j.marpolbul.2011.09.025.

179. Cole S. E. & Stuart K. R. 2000, "Nuclear and cortical histology for brightfield microscopy", in D. J. Asai & J. D. Forney (eds), Methods in Cell Biology, vol. 62, Academic Press, San Diego, pp. 313–322, ISBN 978-0-12-544164-3.

180. Cotton S. A. 1997, Chemistry of Precious Metals, Blackie Academic & Professional, London, ISBN 978-94-010-7154-3.

181. Cotton S. 2006, Lanthanide and Actinide Chemistry, reprinted with corrections 2007, John Wiley & Sons, Chichester, ISBN 978-0-470-01005-1.

182. Cox P. A. 1997, The elements: Their Origin, Abundance and Distribution, Oxford University Press, Oxford, ISBN 978-0-19-855298-7.

183. Crundwell F. K., Moats M. S., Ramachandran V., Robinson T. G. & Davenport W. G. 2011, Extractive Metallurgy of Nickel, Cobalt and Platinum Group Metals, Elsevier, Kidlington, Oxford, ISBN 978-0-08-096809-4.

184. Cui X-Y., Li S-W., Zhang S-J., Fan Y-Y., Ma L. Q. 2015, "Toxic metals in children's toys and jewelry: Coupling bioaccessibility with risk assessment", Environmental Pollution, vol. 200, pp. 77–84, doi:10.1016/j.envpol.2015.01.035.

185. Dapena J. & Teves M. A. 1982, "Influence of the diameter of the hammer head on the distance of a hammer throw", Research Quarterly for Exercise and Sport, vol. 53, no. 1, pp. 78–81, doi:10.1080/02701367.1982.10605229.

186. De Zuane J. 1997, Handbook of Drinking Water Quality, 2nd ed., John Wiley & Sons, New York, ISBN 978-0-471-28789-6.

187. Department of the Navy 2009, Gulf of Alaska Navy Training Activities: Draft Environmental Impact Statement/Overseas Environmental Impact Statement, U.S. Government, accessed 21 August 2016.

188. Deschlag J. O. 2011, "Nuclear fission", in A. Vértes, S. Nagy, Z. Klencsár, R. G. Lovas, F. Rösch (eds), Handbook of Nuclear Chemistry, 2nd ed., Springer Science+Business Media, Dordrecht, pp. 223–280, ISBN 978-1-4419-0719-6.

189. Desoize B. 2004, "Metals and metal compounds in cancer treatment", Anticancer

Research, vol. 24, no. 3a, pp. 1529–1544, PMID 15274320.

190. Dev N. 2008, ‹Modelling Selenium Fate and Transport in Great Salt Lake Wetlands', PhD dissertation, University of Utah, ProQuest, Ann Arbor, Michigan, ISBN 978-0-549-86542-1.
191. Di Maio V. J. M. 2001, Forensic Pathology, 2nd ed., CRC Press, Boca Raton, ISBN 0-8493-0072-X.
192. Di Maio V. J. M. 2016, Gunshot Wounds: Practical Aspects of Firearms, Ballistics, and Forensic Techniques, 3rd ed., CRC Press, Boca Raton, Florida, ISBN 978-1-4987-2570-5.
193. Duffus J. H. 2002, " ‹Heavy metals'—A meaningless term?", Pure and Applied Chemistry, vol. 74, no. 5, pp. 793–807, doi:10.1351/pac200274050793.
194. Dunn P. 2009, Unusual metals could forge new cancer drugs, University of Warwick, accessed 23 March 2016.
195. Ebbing D. D. & Gammon S. D. 2017, General Chemistry, 11th ed., Cengage Learning, Boston, ISBN 978-1-305-58034-3.
196. Edelstein N. M., Fuger J., Katz J. L. & Morss L. R. 2010, "Summary and comparison of properties of the actinde and transactinide elements," in L. R. Morss, N. M. Edelstein & J. Fuger (eds), The Chemistry of the Actinide and Transactinide Elements, 4th ed., vol. 1–6, Springer, Dordrecht, pp. 1753–1835, ISBN 978-94-007-0210-3.
197. Eisler R. 1993, Zinc Hazards to Fish, Wildlife, and Invertebrates: A Synoptic Review, Biological Report 10, U. S. Department of the Interior, Laurel, Maryland, accessed 2 September 2016.
198. Elliott S. B. 1946, The Alkaline-earth and Heavy-metal Soaps, Reinhold Publishing Corporation, New York.
199. Emsley J. 2011, Nature's Building Blocks, new edition, Oxford University Press, Oxford, ISBN 978-0-19-960563-7.
200. Everts S. 2016, "What chemicals are in your tattoo", Chemical & Engineering News, vol. 94, no. 33, pp. 24–26.
201. Fournier J. 1976, "Bonding and the electronic structure of the actinide metals," Journal of Physics and Chemistry of Solids, vol 37, no. 2, pp. 235–244, doi:10.1016/0022-3697(76)90167-0.
202. Frick J. P. (ed.) 2000, Woldman's Engineering Alloys, 9th ed., ASM International, Materials Park, Ohio, ISBN 978-0-87170-691-1.
203. Frommer H. H. & Stabulas-Savage J. J. 2014, Radiology for the Dental Professional, 9th ed., Mosby Inc., St. Louis, Missouri, ISBN 978-0-323-06401-9.
204. Gidding J. C. 1973, Chemistry, Man, and Environmental Change: An Integrated Approach, Canfield Press, New York, ISBN 978-0-06-382790-5.
205. Gmelin L. 1849, Hand-book of chemistry, vol. III, Metals, translated from the German by H. Watts, Cavendish Society, London.
206. Goldsmith R. H. 1982, "Metalloids", Journal of Chemical Education, vol. 59,

no. 6, pp. 526–527, doi:10.1021/ed059p526.

207. Gorbachev V. M., Zamyatnin Y. S. & Lbov A. A. 1980, Nuclear Reactions in Heavy Elements: A Data Handbook, Pergamon Press, Oxford, ISBN 978-0-08-023595-0.

208. Gordh G. & Headrick D. 2003, A Dictionary of Entomology, CABI Publishing, Wallingford, ISBN 978-0-85199-655-4.

209. Greenberg B. R. & Patterson D. 2008, Art in Chemistry; Chemistry in Art, 2nd ed., Teachers Ideas Press, Westport, Connecticut, ISBN 978-1-59158-309-7.

210. Gribbon J. 2016, 13.8: The Quest to Find the True Age of the Universe and the Theory of Everything, Yale University Press, New Haven, ISBN 978-0-300-21827-5.

211. Gschneidner Jr., K. A. 1975, Inorganic compounds, in C. T. Horowitz (ed.), Scandium: Its Occurrence, Chemistry, Physics, Metallurgy, Biology and Technology, Academic Press, London, pp. 152–251, ISBN 978-0-12-355850-3.

212. Guandalini G. S., Zhang L., Fornero E., Centeno J. A., Mokashi V. P., Ortiz P. A., Stockelman M. D., Osterburg A. R. & Chapman G. G. 2011, "Tissue distribution of tungsten in mice following oral exposure to sodium tungstate," Chemical Research in Toxicology, vol. 24, no. 4, pp 488–493, doi:10.1021/tx200011k.

213. Guney M. & Zagury G. J. 2012, "Heavy metals in toys and low-cost jewelry: Critical review of U.S. and Canadian legislations and recommendations for testing", Environmental Science & Technology, vol. 48, pp. 1238–1246, doi:10.1021/es4036122.

214. Habashi F. 2009, "Gmelin and his Handbuch", Bulletin for the History of Chemistry, vol. 34, no. 1, pp. 30–1.

215. Hadhazy A. 2016, "Galactic ‹gold mine' explains the origin of nature's heaviest elements", Science Spotlights, 10 May 2016, accessed 11 July 2016.

216. Hartmann W. K. 2005, Moons & Planets, 5th ed., Thomson Brooks/Cole, Belmont, California, ISBN 978-0-534-49393-6.

217. Harvey P. J., Handley H. K. & Taylor M. P. 2015, "Identification of the sources of metal (lead) contamination in drinking waters in north-eastern Tasmania using lead isotopic compositions," Environmental Science and Pollution Research, vol. 22, no. 16, pp. 12276–12288, doi:10.1007/s11356-015-4349-2 PMID 25895456.

218. Hasan S. E. 1996, Geology and Hazardous Waste Management, Prentice Hall, Upper Saddle River, New Jersey, ISBN 978-0-02-351682-5.

219. Hawkes S. J. 1997, "What is a "heavy metal"?", Journal of Chemical Education, vol. 74, no. 11, p. 1374, doi:10.1021/ed074p1374.

220. Haynes W. M. 2015, CRC Handbook of Chemistry and Physics, 96th ed., CRC Press, Boca Raton, Florida, ISBN 978-1-4822-6097-7.

221. Hendrickson D. J. 2916, "Effects of early experience on brain and body", in D. Alicata, N. N. Jacobs, A. Guerrero and M. Piasecki (eds), Problem-based Behavioural Science and Psychiatry 2nd ed., Springer, Cham, pp. 33–54, ISBN

978-3-319-23669-8.

222. Hermann A., Hoffmann R. & Ashcroft N. W. 2013, "Condensed astatine: Monatomic and metallic", Physical Review Letters, vol. 111, pp. 11604–1−11604-5, doi:10.1103/PhysRevLett.111.116404.

223. Herron N. 2000, "Cadmium compounds," in Kirk-Othmer Encyclopedia of Chemical Technology, vol. 4, John Wiley & Sons, New York, pp. 507–523, ISBN 978-0-471-23896-6.

224. Hoffman D. C., Lee D. M. & Pershina V. 2011, "Transactinide elements and future elements," in L. R. Morss, N. Edelstein, J. Fuger & J. J. Katz (eds), The Chemistry of the Actinide and Transactinide Elements, 4th ed., vol. 3, Springer, Dordrecht, pp. 1652–1752, ISBN 978-94-007-0210-3.

225. Hofmann S. 2002, On Beyond Uranium: Journey to the End of the Periodic Table, Taylor & Francis, London, ISBN 978-0-415-28495-0.

226. Housecroft J. E. 2008, Inorganic Chemistry, Elsevier, Burlington, Massachusetts, ISBN 978-0-12-356786-4.

227. Howell N., Lavers J., Paterson D., Garrett R. & Banati R. 2012, Trace metal distribution in feathers from migratory, pelagic birds, Australian Nuclear Science and Technology Organisation, accessed 3 May 2014.

228. Hübner R., Astin K. B. & Herbert R. J. H. 2010, " ‹Heavy metal'—time to move on from semantics to pragmatics?", Journal of Environmental Monitoring, vol. 12, pp. 1511–1514, doi:10.1039/C0EM00056F.

229. Ikehata K., Jin Y., Maleky N. & Lin A. 2015, "Heavy metal pollution in water resources in China—Occurrence and public health implications", in S. K. Sharma (ed.), Heavy Metals in Water: Presence, Removal and Safety,Royal Society of Chemistry, Cambridge, pp. 141–167, ISBN 978-1-84973-885-9.

230. International Antimony Association 2016, Antimony compounds, accessed 2 September 2016.

231. International Platinum Group Metals Association n.d., The Primary Production of Platinum Group Metals (PGMs), accessed 4 September 2016.

232. Ismail A. F., Khulbe K. & Matsuura T. 2015, Gas Separation Membranes: Polymeric and Inorganic, Springer, Cham, Switzerland, ISBN 978-3-319-01095-3.

233. IUPAC 2016, "IUPAC is naming the four new elements nihonium, moscovium, tennessine, and oganesson" accessed 27 August 2016.

234. Iyengar G. V. 1998, "Reevaluation of the trace element content in Reference Man", Radiation Physics and Chemistry, vol. 51, nos 4–6, pp. 545–560, doi:10.1016/S0969-806X(97)00202-8

235. Jackson J. & Summitt J. 2006, The Modern Guide to Golf Clubmaking: The Principles and Techniques of Component Golf Club Assembly and Alteration, 5th ed., Hireko Trading Company, City of Industry, California, ISBN 978-0-9619413-0-7.

236. Järup L 2003, "Hazards of heavy metal contamination", British Medical Bulletin,

vol. 68, no. 1, pp. 167–182, doi:10.1093/bmb/ldg032.

237. Jones C. J. 2001, d- and f-Block Chemistry, Royal Society of Chemistry, Cambridge, ISBN 978-0-85404-637-9.

238. Kantra S. 2001, "What's new", Popular Science, vol. 254, no. 4, April, p. 10.

239. Keller C., Wolf W. & Shani J. 2012, "Radionuclides, 2. Radioactive elements and artificial radionuclides", in F. Ullmann (ed.), Ullmann's Encyclopedia of Industrial Chemistry, vol. 31, Wiley-VCH, Weinheim, pp. 89–117, doi:10.1002/14356007.o22_o15.

240. King R. B. 1995, Inorganic Chemistry of Main Group Elements, Wiley-VCH, New York, ISBN 978-1-56081-679-9.

241. Kolthoff I. M. & Elving P. J. FR 1964, Treatise on Analytical Chemistry, part II, vol. 6, Interscience Encyclopedia, New York, ISBN 978-0-07-038685-3.

242. Korenman I. M. 1959, "Regularities in properties of thallium", Journal of General Chemistry of the USSR, English translation, Consultants Bureau, New York, vol. 29, no. 2, pp. 1366–90, ISSN 0022-1279.

243. Kozin L. F. & Hansen S. C. 2013, Mercury Handbook: Chemistry, Applications and Environmental Impact, RSC Publishing, Cambridge, ISBN 978-1-84973-409-7.

244. Kumar R., Srivastava P. K., Srivastava S. P. 1994, "Leaching of heavy metals (Cr, Fe, and Ni) from stainless steel utensils in food simulates and food materials", Bulletin of Environmental Contamination and Toxicology, vol. 53, no. 2, doi:10.1007/BF00192942, pp. 259–266.

245. Lach K., Steer B., Gorbunov B., Mička V. & Muir R. B. 2015, "Evaluation of exposure to airborne heavy metals at gun shooting ranges", The Annals of Occupational Hygiene, vol. 59, no. 3, pp. 307–323, doi:10.1093/annhyg/meu097.

246. Landis W., Sofield R. & Yu M-H. 2010, Introduction to Environmental Toxicology: Molecular Substructures to Ecological Landscapes, 4th ed., CRC Press, Boca Raton, Florida, ISBN 978-1-4398-0411-7.

247. Lane T. W., Saito M. A., George G. N., Pickering, I. J., Prince R. C. & Morel F. M. M. 2005, "Biochemistry: A cadmium enzyme from a marine diatom", Nature, vol. 435, no. 7038, p. 42, doi:10.1038/435042a.

248. Lee J. D. 1996, Concise Inorganic Chemistry, 5th ed., Blackwell Science, Oxford, ISBN 978-0-632-05293-6.

249. Leeper G. W. 1978, Managing the Heavy Metals on the Land Marcel Dekker, New York, ISBN 0-8247-6661-X.

250. Lemony A. D. 1997, "A teratogenic deformity index for evaluating impacts of selenium on fish populations", Ecotoxicology and Environmental Safety, vol. 37, no. 3, pp. 259–266, doi:10.1006/eesa.1997.1554.

251. Lide D. R. (ed.) 2004, CRC Handbook of Chemistry and Physics, 85th ed., CRC Press, Boca Raton, Florida, ISBN 978-0-8493-0485-9.

252. Liens J. 2010, "Heavy metals as pollutants", in B. Warf (ed.), Encyclopaedia

of Geography, Sage Publications, Thousand Oaks, California, pp. 1415–1418, ISBN 978-1-4129-5697-0.

253. Lima E., Guerra R., Lara V. & Guzmán A. 2013, "Gold nanoparticles as efficient antimicrobial agents for Escherichia coli and Salmonella typhi " Chemistry Central, vol. 7:11, doi:10.1186/1752-153X-7-11 PMID 23331621 PMC 3556127.

254. Litasov K. D. & Shatskiy A. F. 2016, "Composition of the Earth's core: A review", Russian Geology and Geophysics, vol. 57, no. 1, pp. 22–46, doi:10.1016/j.rgg.2016.01.003.

255. Livesey A. 2012, Advanced Motorsport Engineering, Routledge, London, ISBN 978-0-7506-8908-3.

256. Livingston R. A. 1991, "Influence of the Environment on the Patina of the Statue of Liberty", Environmental Science & Technology, vol. 25, no. 8, pp. 1400–1408, doi:10.1021/es00020a006.

257. Longo F. R. 1974, General Chemistry: Interaction of Matter, Energy, and Man, McGraw-Hill, New York, ISBN 978-0-07-038685-3.

258. Love M. 1998, Phasing Out Lead from Gasoline: Worldwide Experience and Policy Implications, World Bank Technical Paper volume 397, The World Bank, Washington DC, ISBN 0-8213-4157-X.

259. Lyman W. J. 1995, "Transport and transformation processes", in Fundamentals of Aquatic Toxicology, G. M. Rand (ed.), Taylor & Francis, London, pp. 449–492, ISBN 978-1-56032-090-6.

260. Macintyre J. E. 1994, Dictionary of inorganic compounds, supplement 2, Dictionary of Inorganic Compounds, vol. 7, Chapman & Hall, London, ISBN 978-0-412-49100-9.

261. MacKay K. M., MacKay R. A. & Henderson W. 2002, Introduction to Modern Inorganic Chemistry, 6th ed., Nelson Thornes, Cheltenham, ISBN 978-0-7487-6420-4.

262. Magee R. J. 1969, Steps to Atomic Power, Cheshire for La Trobe University, Melbourne.

263. Magill F. N. I (ed.) 1992, Magill's Survey of Science, Physical Science series, vol. 3, Salem Press, Pasadena, ISBN 978-0-89356-621-0.

264. Martin M. H. & Coughtrey P. J. 1982, Biological Monitoring of Heavy Metal Pollution, Applied Science Publishers, London, ISBN 978-0-85334-136-9.

265. Massarani M. 2015, "Brazilian mine disaster releases dangerous metals," Chemistry World, November 2015, accessed 16 April 2016.

266. Masters C. 1981, Homogenous Transition-metal Catalysis: A Gentle Art, Chapman and Hall, London, ISBN 978-0-412-22110-1.

267. Matyi R. J. & Baboian R. 1986, "An X-ray Diffraction Analysis of the Patina of the Statue of Liberty", Powder Diffraction, vol. 1, no. 4, pp. 299–304, doi:10.1017/S0885715600011970.

268. McColm I. J. 1994, Dictionary of Ceramic Science and Engineering, 2nd ed.,

Springer Science+Business Media, New York, ISBN 978-1-4419-3235-8.

269. McCurdy R. M. 1975, Qualities and quantities: Preparation for College Chemistry, Harcourt Brace Jovanovich, New York, ISBN 978-0-15-574100-3.

270. McLemore V. T. (ed.) 2008, Basics of Metal Mining Influenced Water, vol. 1, Society for Mining, Metallurgy, and Exploration, Littleton, Colorado, ISBN 978-0-87335-259-8.

271. McQueen K. G. 2009, Regolith geochemistry, in K. M. Scott & C. F. Pain (eds), Regolith Science, CSIRO Publishing, Collingwood, Victoria, ISBN 978-0-643-09396-6.

272. Mellor J. W. 1924, A comprehensive Treatise on Inorganic and Theoretical Chemistry, vol. 5, Longmans, Green and Company, London.

273. Moore J. W. & Ramamoorthy S. 1984, Heavy Metals in Natural Waters: Applied Monitoring and Impact Assessment, Springer Verlag, New York, ISBN 978-1-4612-9739-0.

274. Morris C. G. 1992, Academic Press Dictionary of Science and Technology, Harcourt Brace Jovanovich, San Diego, ISBN 978-0-12-200400-1.

275. Morstein J. H. 2005, "Fat Man", in E. A. Croddy & Y. Y. Wirtz (eds), Weapons of Mass Destruction: An Encyclopedia of Worldwide Policy, Technology, and History, ABC-CLIO, Santa Barbara, California, ISBN 978-1-85109-495-0.

276. Moselle B. (ed.) 2005, 2004 National Home Improvement Estimator, Craftsman Book Company, Carlsbad, California, ISBN 978-1-57218-150-2.

277. Naja G. M. & Volesky B. 2009, "Toxicity and sources of Pb, Cd, Hg, Cr, As, and radionuclides", in L. K. Wang, J. P. Chen, Y. Hung & N. K. Shammas, Heavy Metals in the Environment, CRC Press, Boca Raton, Florida, ISBN 978-1-4200-7316-4.

278. Nakbanpote W., Meesungneon O. & Prasad M. N. V. 2016, "Potential of ornamental plants for phytoremediation of heavy metals and income generation", in M. N. V. Prasad (ed.), Bioremediation and Bioeconomy, Elsevier, Amsterdam, pp. 179–218, ISBN 978-0-12-802830-8.

279. Nathans M. W. 1963, Elementary Chemistry, Prentice Hall, Englewood Cliffs, New Jersey.

280. National Materials Advisory Board 1971, Trends in the Use of Depleted Uranium, National Academy of Sciences – National Academy of Engineering, Washington DC.

281. National Materials Advisory Board 1973, Trends in Usage of Tungsten, National Academy of Sciences – National Academy of Engineering, Washington DC.

282. National Organization for Rare Disorders 2015, Heavy metal poisoning, accessed 3 March 2016.

283. Natural Resources Canada 2015, "Generation of the Earth's magnetic field", accessed 30 August 2016.

284. Nieboer E. & Richardson D. 1978, "Lichens and ‹heavy metals' ", International

Lichenology Newsletter, vol. 11, no. 1, pp. 1–3.

285. Nieboer E. & Richardson D. H. S. 1980, "The replacement of the nondescript term ‹heavy metals' by a biologically and chemically significant classification of metal ions", Environmental Pollution Series B, Chemical and Physical, vol. 1, no. 1, pp. 3–26, doi:10.1016/0143-148X(80)90017-8.

286. Nzierżanowski K. & Gawroński S. W. 2012, "Heavy metal concentration in plants growing on the vicinity of railroad tracks: a pilot study", Challenges of Modern Technology, vol. 3, no. 1, pp. 42–45, ISSN 2353-4419, accessed 21 August 2016.

287. Ohlendorf H. M. 2003, "Ecotoxicology of selenium", in D. J. Hoffman, B. A. Rattner, G. A. Burton & J. Cairns, Handbook of Ecotoxicology, 2nd ed., Lewis Publishers, Boca Raton, pp. 466–491, ISBN 978-1-56670-546-2.

288. Ondreička R., Kortus J. & Ginter E. 1971, "Aluminium, its absorption, distribution, and effects on phosphorus metabolism", in S. C. Skoryna & D. Waldron-Edward (eds), Intestinal Absorption of Metal Ions, Trace Elements and Radionuclides, Pergamon press, Oxford.

289. Ong K. L., Tan T. H. & Cheung W. L. 1997, "Potassium permanganate poisoning—a rare cause of fatal poisoning", Journal of Accident & Emergency Medicine, vol. 14, no. 1, pp. 43–45, PMC 1342846.

290. Oxford English Dictionary 1989, 2nd ed., Oxford University Press, Oxford, ISBN 978-0-19-861213-1.

291. Pacheco-Torgal F., Jalali S. & Fucic A. (eds) 2012, Toxicity of building materials, Woodhead Publishing, Oxford, ISBN 978-0-85709-122-2.

292. Padmanabhan T. 2001, Theoretical Astrophysics, vol. 2, Stars and Stellar Systems, Cambridge University Press, Cambidge, ISBN 978-0-521-56241-6.

293. Pan W. & Dai J. 2015, "ADS based on linear accelerators", in W. Chao & W. Chou (eds), Reviews of accelerator science and technology, vol. 8, Accelerator Applications in Energy and Security, World Scientific, Singapore, pp. 55–76, ISBN 981-3108-89-4.

294. Parish R. V. 1977, The Metallic Elements, Longman, New York, ISBN 978-0-582-44278-8.

295. Perry J. & Vanderklein E. L. Water Quality: Management of a Natural Resource, Blackwell Science, Cambridge, Massachusetts ISBN 0-86542-469-1.

296. Pickering N. C. 1991, The Bowed String: Observations on the Design, Manufacture, Testing and Performance of Strings for Violins, Violas and Cellos, Amereon, Mattituck, New York.

297. Podosek F. A. 2011, "Noble gases", in H. D. Holland & K. K. Turekian (eds), Isotope Geochemistry: From the Treatise on Geochemistry, Elsevier, Amsterdam, pp. 467–492, ISBN 978-0-08-096710-3.

298. Podsiki C. 2008, "Heavy metals, their salts, and other compounds", AIC News, November, special insert, pp. 1–4.

299. Preschel J. July 29, 2005, "Green bullets not so eco-friendly", CBS News,

accessed 18 March 2016.

300. Preuss P. 17 July, 2011, "What keeps the Earth cooking?," Berkeley Lab, accessed 17 July 2016.

301. Prieto C. 2011, The Adventures of a Cello: Revised Edition, with a New Epilogue, University of Texas Press, Austin, ISBN 978-0-292-72393-1

302. Raghuram P., Soma Raju I. V. & Sriramulu J. 2010, "Heavy metals testing in active pharmaceutical ingredients: an alternate approach", Pharmazie, vol. 65, no. 1, pp. 15–18, doi:10.1691/ph.2010.9222.

303. Rainbow P. S. 1991, "The biology of heavy metals in the sea", in J. Rose (ed.), Water and the Environment, Gordon and Breach Science Publishers, Philadelphia, pp. 415–432, ISBN 978-2-88124-747-7.

304. Rand G. M., Wells P. G. & McCarty L. S. 1995, "Introduction to aquatic toxicology", in G. M. Rand (ed.), Fundamentals of Aquatic Toxicology: Effects, Environmental Fate and Risk Assessment, 2nd ed., Taylor & Francis, London, pp. 3–70, ISBN 978-1-56032-090-6.

305. Rankin W. J. 2011, Minerals, Metals and Sustainability: Meeting Future Material Needs, CSIRO Publishing, Collingwood, Victoria, ISBN 978-0-643-09726-1.

306. Rasic-Milutinovic Z. & Jovanovic D. 2013, "Toxic metals", in M. Ferrante, G. Oliveri Conti, Z. Rasic-Milutinovic & D. Jovanovic (eds), Health Effects of Metals and Related Substances in Drinking Water, IWA Publishing, London, ISBN 978-1-68015-557-0.

307. Raymond R. 1984, Out of the Fiery Furnace: The Impact of Metals on the History of Mankind, Macmillan, South Melbourne, ISBN 978-0-333-38024-6.

308. Rebhandl W., Milassin A., Brunner L., Steffan I., Benkö T., Hörmann M., Burschen J. 2007, "In vitro study of ingested coins: Leave them or retrieve them?", Journal of Paediatric Surgery, vol. 42, no. 10, pp. 1729–1734, doi:10.1016/j.jpedsurg.2007.05.031.

309. Rehder D. 2010, Chemistry in Space: From Interstellar Matter to the Origin of Life, Wiley-VCH, Weinheim, ISBN 978-3-527-32689-1.

310. Renner H., Schlamp G., Kleinwächter I., Drost E., Lüchow H. M., Tews P., Panster P., Diehl M., Lang J., Kreuzer T., Knödler A., Starz K. A., Dermann K., Rothaut J., Drieselmann R., Peter C. & Schiele R. 2012, "Platinum Group Metals and compounds", in F. Ullmann (ed.), Ullmann's Encyclopedia of Industrial Chemistry, vol. 28, Wiley-VCH, Weinheim, pp. 317–388, doi:10.1002/14356007.a21_075.

311. Reyes J. W. 2007, Environmental Policy as Social Policy? The Impact of Childhood Lead Exposure on Crime, National Bureau of Economic Research Working Paper 13097, accessed 16 October 2016.

312. Ridpath I. (ed.) 2012, Oxford Dictionary of Astronomy, 2nd ed. rev., Oxford University Press, New York, ISBN 978-0-19-960905-5.

313. Rockhoff H. 2012, America's Economic Way of War: War and the US Economy

from the Spanish–American War to the Persian Gulf War, Cambridge University Press, Cambridge, ISBN 978-0-521-85940-0.

314. Roe J. & Roe M. 1992, "World's coinage uses 24 chemical elements", World Coinage News, vol. 19, no. 4, pp. 24–25; no. 5, pp. 18–19.

315. Russell A. M. & Lee K. L. 2005, Structure–Property Relations in Nonferrous Metals, John Wiley & Sons, Hoboken, New Jersey, ISBN 978-0-471-64952-6.

316. Rusyniak D. E., Arroyo A., Acciani J., Froberg B., Kao L. & Furbee B. 2010, "Heavy metal poisoning: Management of intoxication and antidotes", in A. Luch (ed.), Molecular, Clinical and Environmental Toxicology, vol. 2, Birkhäuser Verlag, Basel, pp. 365–396, ISBN 978-3-7643-8337-4.

317. Ryan J. 2012, Personal Financial Literacy, 2nd ed., South-Western, Mason, Ohio, ISBN 978-0-8400-5829-4.

318. Samsonov G. V. (ed.) 1968, Handbook of the Physicochemical Properties of the Elements, IFI-Plenum, New York, ISBN 978-1-4684-6066-7.

319. Sanders R. 2003, "Radioactive potassium may be major heat source in Earth's core," UCBerkelyNews, 10 December, accessed 17 July 20016.

320. Schweitzer P. A. 2003, Metallic materials: Physical, Mechanical, and Corrosion properties, Marcel Dekker, New York, ISBN 978-0-8247-0878-8.

321. Schweitzer G. K. & Pesterfield L. L. 2010, The Aqueous Chemistry of the Elements, Oxford University Press, Oxford, ISBN 978-0-19-539335-4.

322. Scott R. M. 1989, Chemical Hazards in the Workplace, CRC Press, Boca Raton, Orlando, ISBN 978-0-87371-134-0.

323. Scoullos M. (ed.), Vonkeman G. H., Thornton I. & Makuch Z. 2001, Mercury — Cadmium — Lead Handbook for Sustainable Heavy Metals Policy and Regulation, Kluwer Academic Publishers, Dordrecht, ISBN 978-1-4020-0224-3.

324. Selinger B. 1978, Chemistry in the Market Place, 2nd ed., Australian National University Press, Canberra, ISBN 978-0-7081-0728-7.

325. Seymour R. J. & O'Farrelly J. 2012, "Platinum Group Metals", Kirk-Other Encyclopaedia of Chemical Technology, John Wiley & Sons, New York, doi:10.1002/0471238961.1612012019052513.a01.pub3.

326. Shaw B. P., Sahu S. K. & Mishra R. K. 1999, "Heavy metal induced oxidative damage in terrestrial plants", in M. N. V. Prased (ed.), Heavy Metal Stress in Plants: From Biomolecules to Ecosystems Springer-Verlag, Berlin, ISBN 978-3-540-40131-5.

327. Shedd K. B. 2002, "Tungsten", Minerals Yearbook, United States Geological Survey.

328. Sidgwick N. V. 1950, The Chemical Elements and their Compounds, vol. 1, Oxford University Press, London.

329. Silva R. J. 2010, "Fermium, mendelevium, nobelium, and lawrencium", in L. R. Morss, N. Edelstein & J. Fuger (eds), The Chemistry of the Actinide and Transactinide Elements, vol. 3, 4th ed., Springer, Dordrecht, pp. 1621–1651,

ISBN 978-94-007-0210-3.

330. Spolek G. 2007, “Design and materials in fly fishing”, in A. Subic (ed.), Materials in Sports Equipment, Volume 2, Woodhead Publishing, Abington, Cambridge, pp. 225–247, ISBN 978-1-84569-131-8.

331. Stankovic S. & Stankocic A. R. 2013, “Bioindicators of toxic metals”, in E. Lichtfouse, J. Schwarzbauer, D. Robert 2013, Green materials for energy, products and depollution, Springer, Dordrecht, ISBN 978-94-007-6835-2, pp. 151–228.

332. State Water Control Resources Board 1987, Toxic substances monitoring program, issue 79, part 20 of the Water Quality Monitoring Report, Sacramento, California.

333. Technical Publications 1953, Fire Engineering, vol. 111, p. 235, ISSN 0015-2587.

334. The Minerals, Metals and Materials Society, Light Metals Division 2016, accessed 22 June 2016.

335. The United States Pharmacopeia 1985, 21st revision, The United States Pharmacopeial Convention, Rockville, Maryland, ISBN 978-0-913595-04-6.

336. Thorne P. C. L. & Roberts E. R. 1943, Fritz Ephraim Inorganic Chemistry, 4th ed., Gurney and Jackson, London.

337. Tisza M. 2001, Physical Metallurgy for Engineers, ASM International, Materials Park, Ohio, ISBN 978-0-87170-725-3.

338. Tokar E. J., Boyd W. A., Freedman J. H. & Wales M. P. 2013, “Toxic effects of metals”, in C. D. Klaassen (ed.), Casarett and Doull’s Toxicology: the Basic Science of Poisons, 8th ed., McGraw-Hill Medical, New York, ISBN 978-0-07-176923-5, accessed 9 September 2016 (subscription required).

339. Tomasik P. & Ratajewicz Z. 1985, Pyridine metal complexes, vol. 14, no. 6A, The Chemistry of Heterocyclic Compounds, John Wiley & Sons, New York, ISBN 978-0-471-05073-5.

340. Topp N. E. 1965, The Chemistry of the Rare-earth Elements, Elsevier Publishing Company, Amsterdam.

341. Torrice M. 2016, “How lead ended up in Flint’s tap water,” Chemical & Engineering News, vol. 94, no. 7, pp. 26–27.

342. Tretkoff E. 2006, “March 20, 1800: Volta describes the Electric Battery”, APS News, This Month in Physics History, American Physical Society, accessed 26 August 2016.

343. Uden P. C. 2005, ‹Speciation of Selenium,’ in R. Cornelis, J. Caruso, H. Crews & K. Heumann (eds), Handbook of Elemental Speciation II: Species in the Environment, Food, Medicine and Occupational Health, John Wiley & Sons, Chichester, pp. 346–65, ISBN 978-0-470-85598-0.

344. United States Environmental Protection Agency 1988, Ambient Aquatic Life Water Quality Criteria for Antimony (III), draft, Office of Research and Development, Environmental Research Laboratories, Washington.

345. United States Environmental Protection Agency 2014, Technical Fact Sheet–Tungsten, accessed 27 March 2016.

346. United States Government 2014, Toxic Pollutant List, Code of Federal Regulations, 40 CFR 401.15., accessed 27 March 2016.

347. Valkovic V. 1990, "Origin of trace element requirements by living matter", in B. Gruber & J. H. Yopp (eds), Symmetries in Science IV: Biological and biophysical systems, Plenum Press, New York, pp. 213–242, ISBN 978-1-4612-7884-9.

348. VanGelder K. T. 2014, Fundamentals of Automotive Technology: Principles and Practice, Jones & Bartlett Learning, Burlington MA, ISBN 978-1-4496-7108-2.

349. Venner M., Lessening M., Pankani D. & Strecker E. 2004, Identification of Research Needs Related to Highway Runoff Management, Transportation Research Board, Washington DC, ISBN 978-0-309-08815-2, accessed 21 August 2016.

350. Venugopal B. & Luckey T. D. 1978, Metal Toxicity in Mammals, vol. 2, Plenum Press, New York, ISBN 978-0-306-37177-6.

351. Vernon R. E. 2013, "Which elements are metalloids", Journal of Chemical Education, vol. 90, no. 12, pp. 1703–1707, doi:10.1021/ed3008457.

352. Volesky B. 1990, Biosorption of Heavy Metals, CRC Press, Boca Raton, ISBN 978-0-8493-4917-1.

353. von Gleich A. 2013, "Outlines of a sustainable metals industry", in A. von Gleich, R. U. Ayres & S. Gößling-Reisemann (eds), Sustainable Metals Management, Springer, Dordrecht, pp. 3–40, ISBN 978-1-4020-4007-8.

354. von Zeerleder A. 1949, Technology of Light Metals, Elsevier Publishing Company, New York.

355. Warth A. H. 1956, The Chemistry and Technology of Waxes, Reinhold Publishing Corporation, New York.

356. Weart S. R. 1983, "The discovery of nuclear fission and a nuclear physics paradigm", in W. Shea (ed.), Otto Hahn and the Rise of Nuclear Physics, D. Reidel Publishing Company, Dordrecht, pp. 91–133, ISBN 978-90-277-1584-5.

357. Weber D. J. & Rutula W. A. 2001, "Use of metals as microbicides in preventing infections in healthcare", in Disinfection, Sterilization, and Preservation, 5th ed., S. S. Block (ed.), Lippincott, Williams & Wilkins, Philadelphia, ISBN 978-0-683-30740-5.

358. Weller G. 1976, Cleaning and Preservation of Coins and Medals, S. J. Durst, New York, ISBN 978-0-915262-03-8.

359. White C. 2010, Projectile Dynamics in Sport: Principles and Applications, Routledge, London, ISBN 978-0-415-47331-6.

360. Wiberg N. 2001, Inorganic Chemistry, Academic Press, San Diego, ISBN 978-0-12-352651-9.

361. Wijayawardena M. A. A., Megharaj M. & Naidu R. 2016, "Exposure, toxicity, health impacts and bioavailability of heavy metal mixtures", in D. L. Sparks, Advances in Agronomy, vol. 138, pp. 175–234, Academic Press, London, ISBN 978-0-12-804774-3.

362. Wingerson L. 1986, "America cleans up Liberty", New Scientist, 25 December/1 January 1987, pp. 31–35, accessed 1 October 2016.

363. Wong M. Y., Hedley G. J., Xie G., Kölln L. S, Samuel I. D. W., Pertegaś A., Bolink H. J., Mosman-Colman, E., "Light-emitting electrochemical cells and solution-processed organic light-emitting diodes using small molecule organic thermally activated delayed fluorescence emitters", Chemistry of Materials, vol. 27, no. 19, pp. 6535–6542, doi:10.1021/acs.chemmater.5b03245.

364. Wulfsberg G. 1987, Principles of Descriptive Inorganic Chemistry, Brooks/Cole Publishing Company, Monterey, California, ISBN 978-0-534-07494-4.

365. Wulfsberg G. 2000, Inorganic Chemistry, University Science Books, Sausalito, California, ISBN 978-1-891389-01-6.

366. Yadav J. S., Antony A., Subba Reddy, B. V. 2012, "Bismuth(III) salts as synthetic tools in organic transformations", in T. Ollevier (ed.), Bismuth-mediated Organic Reactions, Topics in Current Chemistry 311, Springer, Heidelberg, ISBN 978-3-642-27238-7.

367. Yang D. J., Jolly W. L. & O'Keefe A. 1977, "Conversion of hydrous germanium(II) oxide to germynyl sesquioxide, (HGe)2O3", ‹Inorganic Chemistry, vol. 16, no. 11, pp. 2980–2982, doi:10.1021/ic50177a070.

368. Yousif N. 2007, Geochemistry of stream sediment from the state of Colorado using NURE data, ETD Collection for the University of Texas, El Paso, paper AAI3273991.

CHAPTER 8

Soil Salinity and Plant Stress

Chapter Contents

INTRODUCTION

Soil salinity is the salt content in the soil; the process of increasing the salt content is known as salinization. Salts occur naturally within soils and water. Salination can be caused by natural processes such as mineral weathering or by the gradual withdrawal of an ocean. It can also come about through artificial processes such as irrigation.

Natural Occurrence

Salts are a natural component in soils and water. The ions responsible for salination are: Na^+, K^+, Ca^{2+}, Mg^{2+} and Cl^-.As the Na^+ (sodium) predominates, soils can become **sodic**. Sodic soils present particular challenges because they tend to have very poor structure which limits or prevents water infiltration and drainage.

Over long periods of time, as soil minerals weather and release salts, these salts are flushed or leached out of the soil by drainage water in areas with sufficient precipitation. In addition to mineral weathering, salts are also deposited via dust and precipitation. In dry regions salts may accumulate, leading to naturally saline soils. This is the case, for example, in large parts of Australia. Human practices can increase the salinity of soils by the addition of salts in irrigation water. Proper irrigation management can prevent salt accumulation by providing adequate drainage water to leach added salts from the soil. Disrupting drainage patterns that provide leaching can also result in salt accumulations. An example of this occurred in Egypt in 1970 when the Aswan High Dam was built. The change in the level of ground water before the construction had enabled soil erosion, which led to high concentration of salts in the water table. After the construction, the continuous high level of the water table led to the salination of the arable land.

Dry Land Salinity

Salinity in drylands can occur when the water table is between two and three metres from the surface of the soil. The salts from the groundwater are raised by capillary action to the surface of the soil. This occurs when groundwater is saline (which is true in many areas), and is favored by land use practices allowing more rainwater to enter the aquifer than it could accommodate. For example, the clearing of trees for agriculture is a major reason for dryland salinity in some areas, since deep rooting of trees has been replaced by shallow rooting of annual crops.

Salinity Due to Irrigation

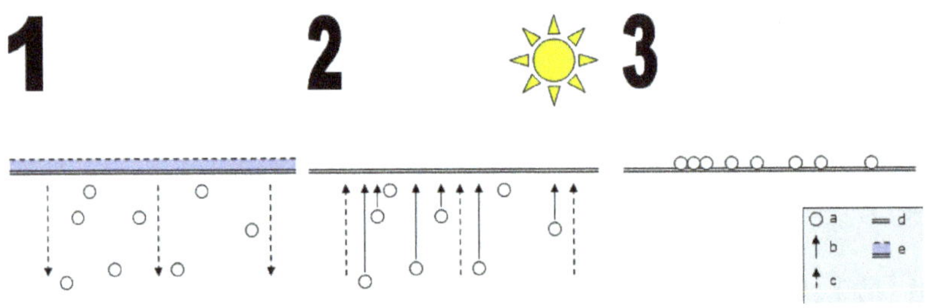

Rain or irrigation, in the absence of leaching, can bring salts to the surface by capillary action

Salinity from irrigation can occur over time wherever irrigation occurs, since almost all water (even natural rainfall) contains some dissolved salts. When the plants use the water, the salts are left behind in the soil and eventually begin to accumulate. Since soil salinity makes it more difficult for plants to absorb soil moisture, these salts must be leached out of the plant root zone by applying additional water. This water in excess of plant needs is called the leaching fraction. Salination from irrigation water is also greatly increased by poor drainage and use of saline water for irrigating agricultural crops.

Salinity in urban areas often results from the combination of irrigation and groundwater processes. Irrigation is also now common in cities (gardens and recreation areas).

Consequences of Salinity

The consequences of salinity are

- detrimental effects on plant growth and yield
- damage to infrastructure (roads, bricks, corrosion of pipes and cables)
- reduction of water quality for users, sedimentation problems
- soil erosion ultimately, when crops are too strongly affected by the amounts of salts.

Salinity is an important land degradation problem. Soil salinity can be reduced by leaching soluble salts out of soil with excess irrigation water. Soil salinity control involves watertable control and flushing in combination with tile drainage or another form of subsurface drainage. A comprehensive treatment of soil salinity is available from the United Nations Food and Agriculture Organization.

High levels of soil salinity can be tolerated if salt-tolerant plants are grown. Sensitive crops lose their vigor already in slightly saline soils, most crops are negatively affected by (moderately) saline soils, and only salinity resistant crops thrive in severely saline soils. The University of Wyoming and the Government of Alberta report data on the salt tolerance of plants.

Regions Affected

From the FAO/UNESCO Soil Map of the World the following salinised areas can be derived.

DRYLAND SALINITY

Dryland salinity is a natural process for soil, just like other processes such as wind erosion. Salinity degrades land by an increase in soil salt concentration in the environment, watercourse or soil in unirrigated landscapes, being in excess of normal soil salt concentrations in dryland regions.

Overview

Salinity refers to the movement and concentration of salt in the landscape and its associated detriment to land and water resources; dryland salinity refers to salinity in unirrigated landscapes. Salinity processes extend from local to regional scales and are driven by imbalances in the water budget that result, primarily, from agriculturally driven landscape change. There are two types of salinity:

Types of Salinity

There are two types of salinity. Primary salinity (natural salinity) and Secondary salinity (induced salinity). (Nrm.qld.gov.au, 2013) Primary salinity naturally occurs in arid and saline environments such as salt lakes, marshes, pans and salt flats. Natural accumulation of salt in soils is an outcome from previous cycles of drainage, desiccation and sea winds. High levels of salt are often found in dry soils, more so than wet soils as it is diluted and washed through the soil profile. (Barry and Holwell et al., 2012).

Secondary salinity is a direct result of human interaction with the land, during development, agriculture and irrigation. Certain land practices have led to changes in the natural structure of the biosphere resulting in excess salting of the land, waterways and soils; thus having detrimental effects on biodiversity and the lands' productivity.

Salinity and Water Cycle

Factors such as climate, features of landscape, soils, drainage, aspect and the effects of human activities; all impact on the severity and occurrence of dryland salinity. Dryland salinity effects human and natural resources, such as native vegetation and crops, animals, infrastructure, agricultural inputs, biodiversity, aquatic ecosystems and water supply quality in the environment. Understanding dryland salinity requires a look at the water cycle. From precipitation, water first enters the soil through a process called infiltration. The water will then remain indefinitely within the spaces or pores between soil particles as soil moisture. Soil moisture may be lost to the surface or atmosphere directly, or through plant uptake, in a process called evapotranspiration. Soil moisture may also continue to move downward to join the groundwater, through a process called groundwater recharge. Recharge is most likely to occur when the amount of water that is available to the soil exceeds the soil's capacity to store it (field capacity). Recharge may also occur by saturated flow when water bypasses the soil matrix as it moves to depth in macropores (e.g. root holes, fractures). Excessive recharge may raise the water table locally, or at a landscape scale. When brackish to saline groundwater intersects the ground surface and discharges, this is termed saline discharge. Areas of discharge are called saline seeps (when groundwater intersects the soil surface) or saline scalds (where water is lost by evaporation only). Groundwater discharge manifests in such problems as: reduced agricultural production, degradation of natural environment, reduced surface water quality, damage to infrastructure including roads, as well as soil erosion and denudation of land.

Reasons for Dryland Salinity

Dryland salinity is broadly the result of three processes: Groundwater recharge, Ground water movement and Groundwater discharge. (Environment.nsw.gov.au, 2013)

Groundwater recharge occurs naturally, being a key process in the development of salinity, however land clearing accelerates this development as the once present deep-rooted plants do not utilize excess runoff, and it now seeps past the roots zone to enter the groundwater system. (Environment.nsw.gov.au, 2013) Evapotranspiration is reduced, as vegetation is lost, resulting in an imbalance in groundwater recharge and discharge, causing the water table to rise. Capillary action brings salt to the surface initiating surface salt accumulation. (Barry and Holwell et al., 2012)

Prevention and alleviation of dry land salinity is a simplistic concept in theory however complex in application. Ceasing the removal of deeply rooted vegetation in order to moderate unbalanced groundwater recharge and the replanting of deeply rooted vegetation such as Eucalyptus and salt tolerant species in regions where salinity is present, will start to alleviate the salt and ground water discharge problems. (Barry and Holwell et al., 2012) However establishing plants in salt laden areas is extremely difficult.

Dryland salinity is a sign that the water balance of the nearby area of land or catchment has been altered. Clearing as little as 25% of a catchment can cause salinity to occur. In addition to adding extra recharge, salinity may also be caused if the aquifers discharge capacity has been exceeded. In many Australian landscapes, aquifer capacity may be several orders of magnitude below that of the altered recharge. Restoring the balance requires either the introduction of natural vegetation (e.g. mallee eucalyptus or perennial grasses), which intercepts and transpires most of the incoming rainfall; or by adapting agriculture to the increased area of shallow, saline groundwater.

Management

Role of Soils

Dryland salinity management often focuses on vegetation, yet it is the collective role of soils and vegetation that has an effect on the root cause of the problem, recharge. Soil health cannot be ignored as a valuable and extensive activity for the management of dryland salinity – the multiple benefits of improving soil health are clear and can be motivated by the potential for local and regional economic and social gains.

Soil is considered in two contexts when it comes to dryland salinity: Recharge and discharge.

Soils in Groundwater Recharge Areas

Soils absorb and store water according to their water holding or field capacity and how dry they are to start with. In much of Victoria, under typical rainfall and natural vegetation cover, soils take on water during wet winters and dry out over summers as plants consume the water (Young & Young, 2001). The drier the soil when leading into winter, the more water can be stored that might otherwise leak to groundwater.

To reduce recharge to levels that existed in pre-clearing conditions is generally infeasible in most agricultural landscapes. This is because there are too few profitable perennials that can replace crops or that can be adopted at the scale required. In most recharge areas that are at risk of future salinity, the goal is to minimise recharge. This can be done by planting wide-spaced trees (alley farming), areas of perennials on suitable soils, and by preventing soils from being left without a significant leaf area in winter and spring. In recharge areas remote from saline areas there is often little incentive for farmers to adopt low profit, low recharge systems. In addition, recharge is the source of fresh groundwater, and a trade off between consumptive use and salinity should be acknowledged.

Soils in Groundwater Discharge Areas

The manifestation of dryland salinity is largely a problem of groundwater – however the accumulation of salt within the soil and at the surface due to proximity to or saturation by saline groundwater causes changes to the soil's chemistry, structure and stability, and the plant life that it supports.

Managing Soils for Dryland Salinity in Catchments

In discharge areas, salinity can be managed by establishing salt-tolerant plants or by engineering systems. Engineering systems include deep open drains, pumps, siphons and various forms of surface water management. Engineering system involve the obvious discharge of salt and water. While saline areas also discharge salt and water, the abundance and timing will be changed. In most Australian states, farmers would be advised to seek advice before using engineering systems.

Establishing salt-tolerant plants can improve salt discharge rates and improve soil health. Improvements undertaken at a catchment scale bring many benefits. Some of these are providing for increased agricultural and associated regional productivity, and utilising water for production that otherwise would contribute to an environmental problem.

SOIL SALINITY CONTROL

Soil salinity control relates to controlling the problem of soil salinity and reclaiming salinized agricultural land.

The aim of soil salinity control is to prevent soil degradation by salination and reclaim already salty (saline) soils. Soil reclamation is also called soil improvement, rehabilitation, remediation, recuperation, or amelioration.

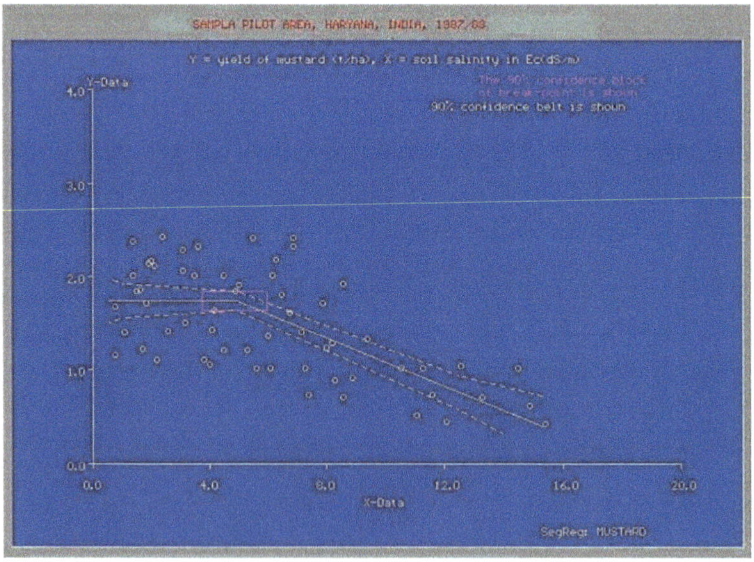

SegReg program: yield of mustard (colza) and soil salinity

The primary man-made cause of salinization is irrigation. River water or groundwater used in irrigation contains salts, which remain behind in the soil after the water has evaporated.

The primary method of controlling soil salinity is to permit 10-20% of the irrigation water to leach the soil, be drained and discharged through an appropriate drainage system. The salt concentration of the drainage water is normally 5 to 10 times higher than that of the irrigation water, thus salt export matches salt import and it will not accumulate.

The Soil Salinity Problem

Salty (saline) soils are soils that have a high salt content. The predominant salt is normally sodium chloride (NaCl, "table salt"). Saline soils are therefore also sodic soils but there may be sodic soils that are not saline, but alkaline.

According to a study by UN University, about 62 million hectares (20%) of the world's irrigated lands are affected, up from 45 million hectares in the early 1990s. In the Indo-Gangetic Plain, home to over 10% of the world's population, crop yield losses for wheat, rice, sugarcane and cotton grown on salt-affected lands could be 40%, 45%, 48%, and 63%, respectively.

Salty soils are a common feature and an environmental problem in irrigated lands in arid and semi-arid regions, resulting in poor or little crop production. The problems are often associated with high water tables, caused by a lack of natural subsurface drainage to the underground. Poor subsurface drainage may be caused by insufficient transport capacity of the aquifer or because water cannot exit the aquifer, for instance if the aquifer is situated in a topographical depression.

Worldwide, the major factor in the development of saline soils is a lack of precipitation. Most naturally saline soils are found in (semi)arid regions and climates of the earth.

Primary Cause

Irrigated saline land with poor crop stand

For example, assuming irrigation water with a low salt concentration of 0.3 g/l (equal to 0.3 kg/m^3 corresponding to an electric conductivity of about 0.5 FdS/m) and a modest annual supply of irrigation water of 10,000 m^3/ha (almost 3 mm/day) brings 3,000 kg salt/ha each year. In the absence of sufficient natural drainage (as in waterlogged soils) and without a proper leaching and drainage program to remove salts, this would lead to a high soil salinity and reduced crop yields in the long run.

Much of the water used in irrigation has a higher salt content than in this example, which is compounded by the fact that many irrigation projects use a far greater annual supply of water. Sugar cane, for example, needs about 20,000 m^3/ha of water per year. As a result, irrigated areas often receive more than 3,000 kg/ha of salt per year and some receive as much as 10,000 kg/ha/year.

Secondary Cause

The secondary cause of salinization is waterlogging in irrigated land. Irrigation causes changes to the natural water balance of irrigated lands. Large quantities of water in irrigation projects are not consumed by plants and must go somewhere. In irrigation projects it is impossible to achieve 100% irrigation efficiency where all the irrigation water is consumed by the plants. The maximum attainable irrigation efficiency is about 70% but usually it is less than 60%. This means that minimum 30%, but usually more than 40% of the irrigation water is not evaporated and it must go somewhere.

Most of the water lost this way is stored underground which can change the original hydrology of local aquifers considerably. Many aquifers cannot absorb and transport these quantities of water and so the water table rises leading to water logging.

Waterlogging causes three problems:

- The shallow water table and lack of oxygenation of the root zone reduces the yield of most crops
- It leads to an accumulation of salts brought in with the irrigation water as their removal through the aquifer is blocked
- With the upward seepage of groundwater more salts are brought into the soil and the salination is aggravated

Aquifer conditions in irrigated land and the groundwater flow have an important role in soil salinization, as illustrated here :

Illustration of the influence of aquifer conditions on soil salinization in irrigated land

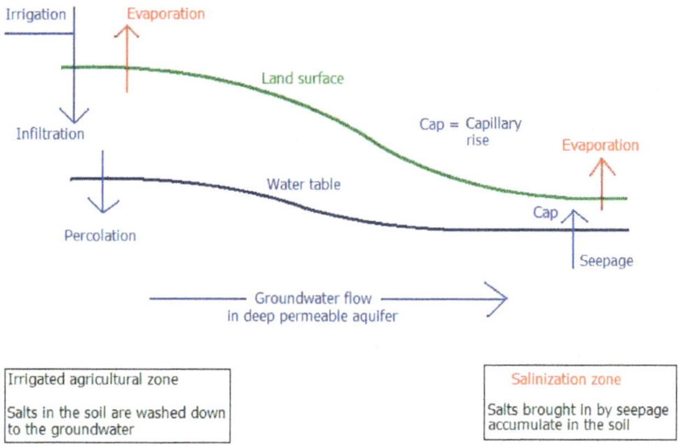

Soil salinization in the lower parts of undulating land with a good aquifer

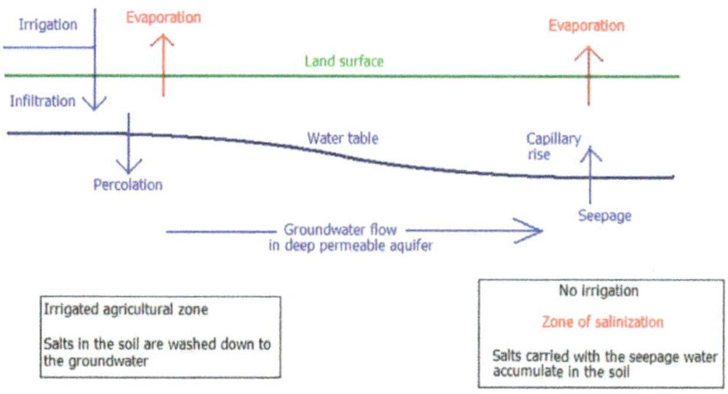

Soil salinization in the unirrigated parts of flat land with a good aquifer

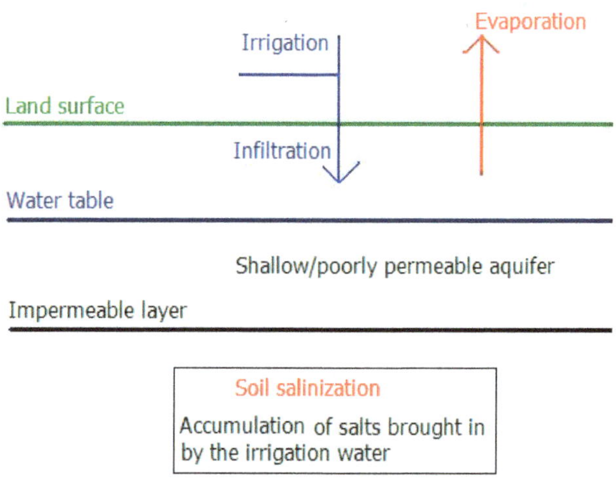

Soil salinization in irrigated flat land without an aquifer

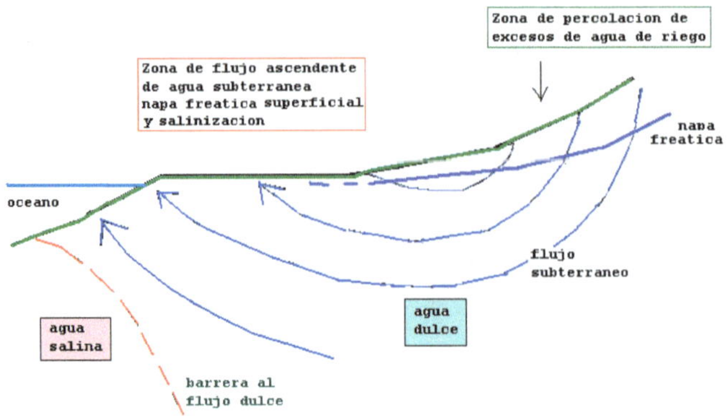

Soil salinization in a coastal delta from irrigation higher up

Salt Affected Area

Normally, the salinization of agricultural land affects a considerable area of irrigation projects, on the order of 20 to 30%. When the agriculture in such a fraction of the land is abandoned, a new salt and water balance is attained, a new equilibrium is reached, and the situation becomes stable.

In India alone, thousands of square kilometres have been severely salinized. China and Pakistan do not lag much behind (perhaps China has even more salt affected land than India). A regional distribution of the 3,230,000 km^2 of saline land worldwide is shown in the following table derived from the FAO/UNESCO Soil Map of the World.

Region	Area (10^6ha)
Australia	84.7
Africa	69.5
Latin America	59.4
Near and Middle East	53.1
Europe	20.7
Asia and Far East	19.5
Northern America	16.0

Spatial Variation

Although the principles of the processes of salinization are fairly easy to understand, it is more difficult to explain why certain parts of the land suffer from the problems and other parts do not, or to predict accurately which part of the land will fall victim. The main reason for this is the variation of natural conditions in time and space, the usually uneven distribution of the irrigation water, and the seasonal or yearly changes of agricultural practices. Only in lands with undulating topography is the prediction simple: the depressional areas will degrade the most.

The preparation of salt and water balances for distinguishable sub-areas in the irrigation project, or the use of agro-hydro-salinity models, can be helpful in explaining or predicting the extent and severity of the problems.

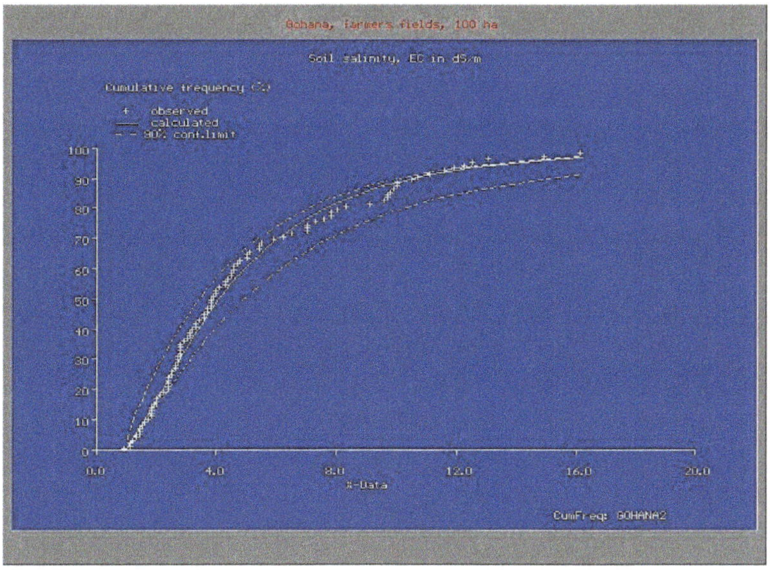

CumFreq program: spatial variation of soil salinity

Diagnosis

Measurement

Soil salinity is measured as the salt concentration of the soil solution in tems of g/l or electric conductivity (EC) in dS/m. The relation between these two units is about 5/3 : y g/l => 5y/3 dS/m. Seawater may have a salt concentration of 30 g/l (3%) and an EC of 50 dS/m.

The standard for the determination of soil salinity is from an extract of a saturated paste of the soil, and the EC is then written as ECe. The extract is obtained by centrifugation. The salinity can more easily be measured, without centrifugation, in a 2:1 or 5:1 water:soil mixture (in terms of g water per g dry soil) than from a saturated paste. The relation between ECe and $EC_{2:1}$ is about 4, hence : ECe = 4 $EC_{1:2}$.

Classification

Soils are considered saline when the ECe > 4. The soil is called slightly saline when 4 < ECe < 8, (moderately) saline when 8 < ECe < 16, and severely saline when ECe > 16.

Crop Tolerance

Sensitive crops lose their vigor already in slightly saline soils, most crops are negatively affected by (moderately) saline soils, and only salinity resistant crops thrive in severely saline soils. The University of Wyoming and the Government of Alberta report data on the salt tolerance of plants.

Principles of Salinity Control

Drainage is the primary method of controlling soil salinity. The system should permit a small fraction of the irrigation water (about 10 to 20 percent, the drainage or leaching fraction) to be drained and discharged out of the irrigation project.

In irrigated areas where salinity is stable, the salt concentration of the drainage water is normally 5 to 10 times higher than that of the irrigation water. Salt export matches salt import and salt will not accumulate.

When reclaiming already salinized soils, the salt concentration of the drainage water will initially be much higher than that of the irrigation water (for example 50 times higher). Salt export will greatly exceed salt import, so that with the same drainage fraction a rapid desalinization occurs. After one or two years, the soil salinity is decreased so much, that the salinity of the drainage water has come down to a normal value and a new, favorable, equilibrium is reached.

In regions with pronounced dry and wet seasons, the drainage system may be operated in the wet season only, and closed during the dry season. This practice of checked or controlled drainage saves irrigation water.

The discharge of salty drainage water may pose *environmental problems* to downstream areas. The environmental hazards must be considered very carefully and, if necessary mitigating measures must be taken. If possible, the drainage must be limited to wet seasons only, when the salty effluent inflicts the least harm.

Drainage Systems

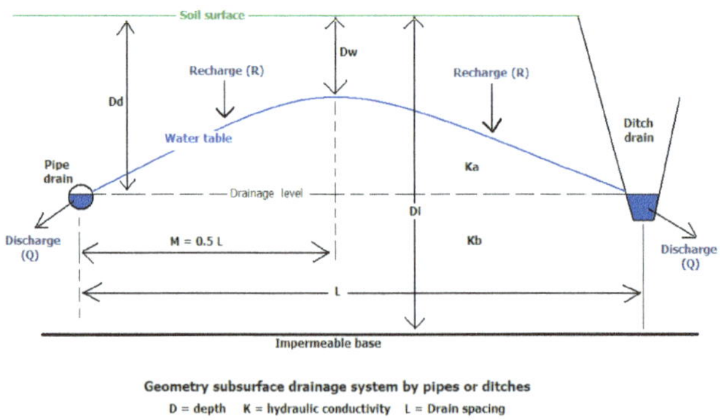

Parameters of a horizontal drainage system

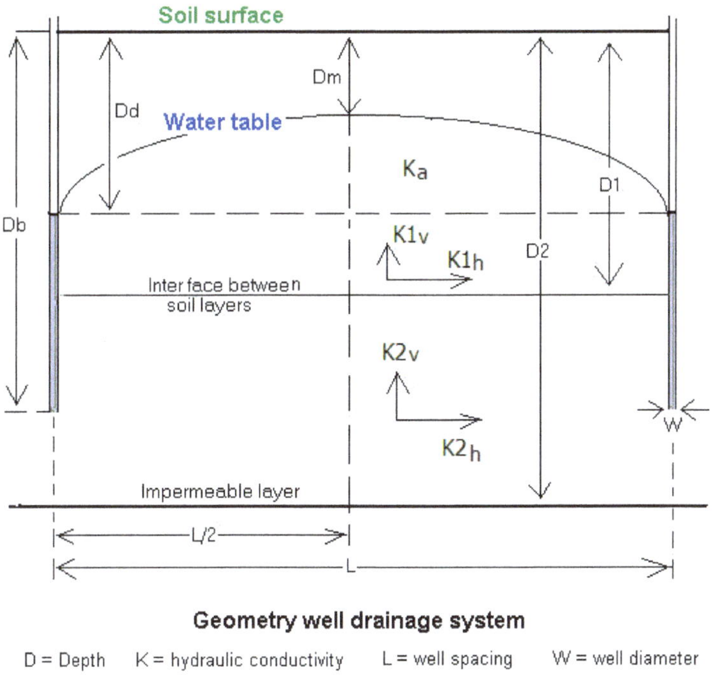

Parameters of a vertical drainage system

Land drainage for soil salinity control is usually by *horizontal drainage system* (figure left), but *vertical systems* (figure right) are also employed.

The drainage system designed to evacuate salty water also lowers the water table. To reduce the cost of the system, the lowering must be reduced to a minimum. The highest permissible level of the water table (or the shallowest permissible depth) depends on the irrigation and agricultural practices and kind of crops.

In many cases a seasonal average water table depth of 0.6 to 0.8 m is deep enough. This means that the water table may occasionally be less than 0.6 m (say 0.2 m just after an irrigation or a rain storm). This automatically implies that, in other occasions, the water table will be deeper than 0.8 m (say 1.2 m). The fluctuation of the water table helps in the breathing function of the soil while the expulsion of carbon dioxide (CO_2) produced by the plant roots and the inhalation of fresh oxygen (O_2) is promoted.

The establishing of a not too deep water table offers the additional advantage that excessive field irrigation is discouraged, as the crop yield would be negatively affected by the resulting elevated water table, and irrigation water may be saved.

The statements made above on the optimum depth of the water table are very general, because in some instances the required water table may be still shallower than indicated (for example in rice paddies), while in other instances it must be considerably deeper (for example in some orchards). The establishment of the optimum depth of the water table is in the realm of agricultural drainage criteria.

Soil Leaching

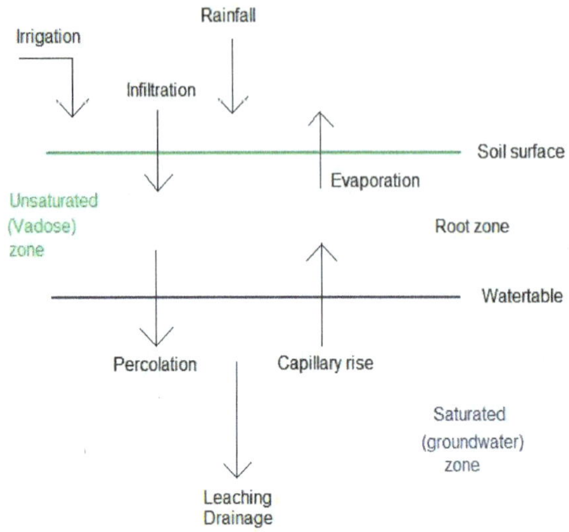

Water balance factors in the soil

The vadose zone of the soil below the soil surface and the watertable is subject to four main hydrological inflow and outflow factors:

- Infiltration of rain and irrigation water (Irr) into the soil through the soil surface (Inf) :
- Inf = Rain + Irr
- Evaporation of soil water through plants and directly into the air through the soil surface (Evap)
- Percolation of water from the unsaturated zone soil into the groundwater through the watertable (Perc)
- Capillary rise of groundwater moving by capillary suction forces into the unsaturated zone(Cap)

In steady state (i.e. the amount of water stored in the unsaturated zone does not change in the long run) the water balance of the unsaturated zone reads: Inflow = Outflow, thus:

- Inf + Cap = Evap + Perc or :
- Irr + Rain + Cap = Evap + Perc

and the *salt balance* is

- Irr.Ci + Cap.Cc = Evap.Fc.Ce + Perc.Cp + Ss

where Ci is the salt concentration of the irrigation water, Cc is the salt concentration of the capillary rise, equal to the salt concentration of the upper part of the groundwater body, Fc is the fraction of the total evaporation transpired by plants, Ce is the salt concentration of the water taken up by the plant roots, Cp is the salt concentration of the percolation water, and Ss is the increase of salt storage in the unsaturated soil. This assumes that the rainfall contains no salts. Only along the coast this may not be true. Further it is assumed that no runoff or surface drainage occurs. The amount of removed by plants (Evap.Fc.Ce) is usually negligibly small: Evap.Fc.Ce = 0

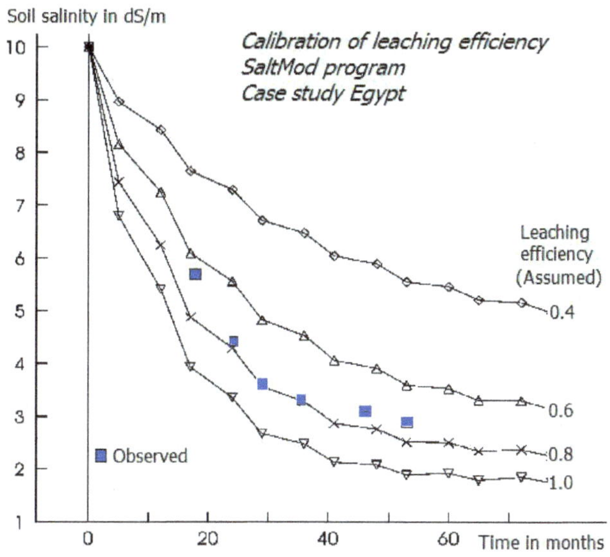

Leaching curves, calibrating leaching efficiency

The salt concentration Cp can be taken as a part of the salt concentration of the soil in the unsaturated zone (Cu) giving: Cp=Le.Cu, where Le is the leaching efficiency. The leaching efficiency is often in the order of 0.7 to 0.8, but in poorly structured, heavy clay soils it may be less. In the Leziria Grande polder in the delta of the Tagus river in Portugal it was found that the leaching efficiency was only 0.15. Assuming that one wishes to avoid the soil salinity to increase and maintain the soil salinity Cu at a desired level Cd we have: Ss = 0, Cu = Cd and Cp = Le.Cd. Hence the salt balance can be simplified to:

- Perc.Le.Cd = Irr.Ci + Cap.Cc

Setting the amount percolation water required to fulfill this salt balance equal to Lr (the *leaching requirement*) it is found that:

- Lr = (Irr.Ci + Cap.Cc) / Le.Cd .

Substituting herein Irr = Evap + Perc − Rain − Cap and re-arranging gives :

- Lr = [(Evap−Rain).Ci + Cap(Cc−Ci)] / (Le.Cd − Ci)

With this the irrigation and drainage requirements for salinity control can be computed too. In irrigation projects in (semi)arid zones and climates it is important to check the leaching requirement, whereby the *field irrigation efficiency* (indicating the fraction of irrigation water percolating to the underground) is to be taken into account. The desired soil salinity level Cd depends on the crop tolerance to salt. The University of Wyoming, USA, and the Government of Alberta,Canada, report crop tolerance data.

Strip Cropping: an Alternative

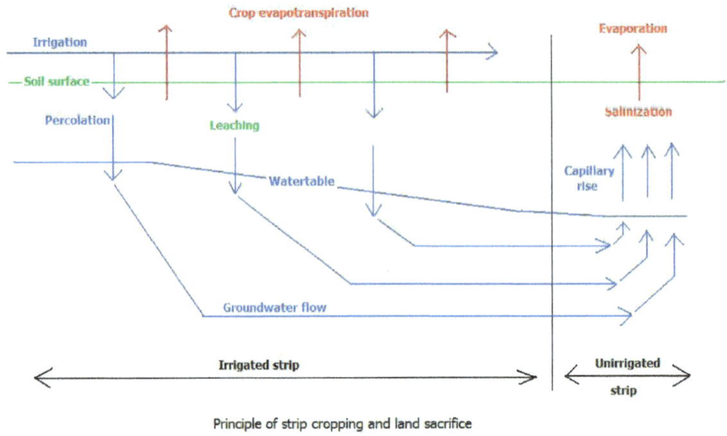

Hydrological principles of *strip cropping* to control the depth of the water table and the soil salinity

In irrigated lands with scarce water resources suffering from drainage (high water table) and soil salinity problems, strip cropping is sometimes practiced with strips of land where every other strip is irrigated while the strips in between are left permanently fallow.

Owing to the water application in the irrigated strips they have a higher watertable which induces flow of groundwater to the unirrigated strips. This flow functions as subsurface drainage for the irrigated strips, whereby the water table is maintained at a not-too-shallow depth, leaching of the soil is possible, and the soil salinity can be controlled at an acceptably low level.

In the unirrigated (sacrificial) strips the soil is dry and the groundwater comes up by capillary rise and evaporates leaving the salts behind, so that here the soil salinizes. Nevertheless, they can have some use for livestock, sowing salinity resistant grasses or weeds. Moreover, useful salt resistant trees can be planted like Casuarina, Eucalyptus or Atriplex, keeping in mind that the trees have deep rooting systems and the salinity of the wet subsoil is less than of the topsoil. In these ways wind erosion can be controlled. The unirrigated strips can also be used for salt harvesting.

Soil Salinity Models

The majority of the computer models available for water and solute transport in the soil (e.g. SWAP,DrainMod-S, UnSatChem, and Hydrus) are based on Richard's differential equation for the movement of water in unsaturated soil in combination with Fick's differential convection–diffusion equation for advection and dispersion of salts.

The models require input of soil characteristics like the relations between variable unsaturated soil moisture content, water tension, water retention curve, unsaturated hydraulic conductivity, dispersivity and diffusivity. These relations vary to a great extent from place to place and from time to time and are not easy to measure. Further, the models are difficult to calibrate under farmer's field conditions because the soil salinity here is spatially very variable. The models use short time steps and need at least a daily, if not an hourly, data base of hydrological phenomena. Altogether this makes model application to a fairly large project the job of a team of specialists with ample facilities.

Simpler models, like SaltMod, based on monthly or seasonal water and soil balances and an empirical capillary rise function, are also available. They are useful for long-term salinity predictions in relation to irrigation and drainage practices.

LeachMod, using the SaltMod principles, helps in analyzing leaching experiments in which the soil salinity was monitored in various root zone layers. Meanwhile, the model will optimize the value of the leaching efficiency of each layer so that a fit, observed with simulated soil salinity values, is obtained.

Spatial variations owing to variations in topography can be simulated and predicted using salinity cum groundwater models, like SahysMod.

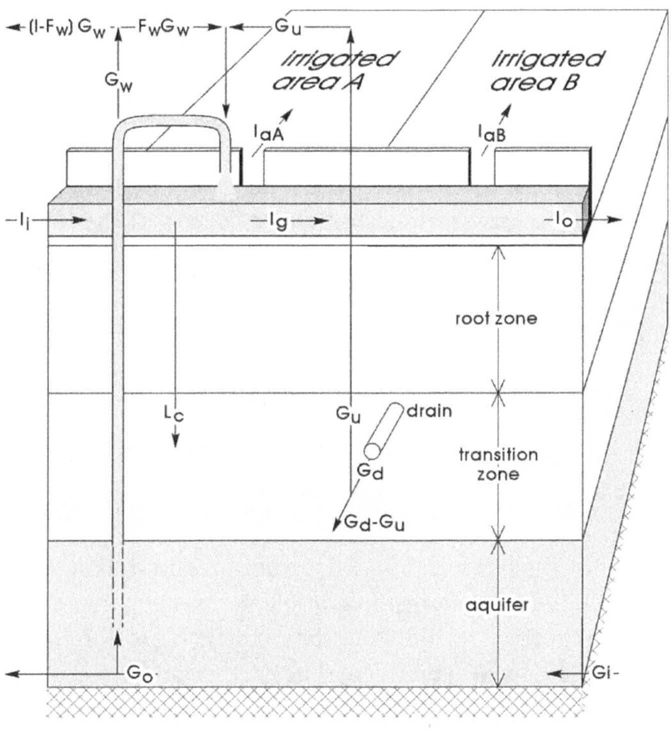

SaltMod components

REFERENCES

1. from "Soil salinity" in WaterWiki, the on-line Knowledge and Collaboration Tool of the Community of Practice (CoP) on Water- and UNDP-related activities in Central and South-Eastern Europe, Caucasus and Central Asia.

2. ILRI (1989), Effectiveness and Social/Environmental Impacts of Irrigation Projects: a Review (PDF), In: Annual Report 1988 of the International Institute for Land Reclamation and Improvement (ILRI), Wageningen, The Netherlands, pp. 18–34
3. Drainage Manual: A Guide to Integrating Plant, Soil, and Water Relationships for Drainage of Irrigated Lands, Interior Dept., Bureau of Reclamation, 1993, ISBN 0-16-061623-9
4. “Free articles and software on drainage of waterlogged land and soil salinity control”. Retrieved 2010-07-28.
5. Salt-Affected Soils and their Management, FAO Soils Bulletin 39 (http://www.fao.org/docrep/x5871e/x5871e00.htm)
6. Alan D. Blaylock, 1994, Soil Salinity and Salt tolerance of Horticultural and Landscape Plants. University of Wyoming
7. Government of Alberta, Salt tolerance of Plants
8. R.Brinkman, 1980. Saline and sodic soils. In: Land reclamation and water management, p. 62-68. International Institute for Land Reclamation and Improvement (ILRI), Wageningen, The Netherlands.
9. http://phys.org/news/2014-10-world-hectares-farm-soil-daily.html
10. I.P. Abrol, J.S.P Yadav, and F. Massoud 1988. Salt affected soils and their management, Food and Agricultural Organization of the United Nations (FAO), Soils Bulletin 39.
11. ILRI, 2003. Drainage for Agriculture: Drainage and hydrology/salinity - water and salt balances. Lecture notes International Course on Land Drainage, International Institute for Land Reclamation and Improvement (ILRI), Wageningen, The Netherlands. Download from page : , or directly as PDF :
12. R.Brinkman, 1980. Saline and sodic soils. In: Land reclamation and water management, p. 62-68. International Institute for Land Reclamation and Improvement (ILRI), Wageningen, The Netherlands.
13. SaltMod: a tool for interweaving of irrigation and drainage for salinity control. In: W.B.Snellen (ed.), Towards integration of irrigation, and drainage management. ILRI Special report, p. 41-43. International Institute for Land Reclamation and Improvement (ILRI), Wageningen, The Netherlands.
14. ILRI, 2003, This paper discusses soil salinity. Lecture notes, International Course on Land Drainage International Institute for Land Reclamation and Improvement (ILRI), Wageningen, The Netherlands. On line:
15. L.A.Richards (Ed.), 1954. Diagnosis and improvement of saline and alkali soils. USDA Agricultural Handbook 60. On internet
16. Alan D. Blaylock, 1994, Soil Salinity and Salt tolerance of Horticultural and Landscape Plants.
17. Government of Alberta, Salt tolerance of Plants

18. J.W. van Hoorn and J.G. van Alphen (2006), Salinity control. In: H.P. Ritzema (Ed.), Drainage Principles and Applications, p. 533-600, Publication 16, International Institute for Land Reclamation and Improvement (ILRI), Wageningen, The Netherlands. ISBN 90-70754-33-9.
19. Agricultural Drainage Criteria, Chapter 17 in: H.P.Ritzema (2006), Drainage Principles and Applications, Publication 16, International Institute for Land Reclamation and Improvement (ILRI), Wageningen, The Netherlands. ISBN 90-70754-33-9. On line :
20. R.J.Oosterbaan and M.A.Senna, 1990. Using SaltMod to predict drainage and salinity control in the Nile delta. In: Annual Report 1989, International Institute for Land Reclamation and Improvement (ILRI), Wageningen, The Netherlands, p. 63-74. See Case Study Egypt in the SaltMod manual :
21. E.A. Vanegas Chacon, 1990. Using SaltMod to predict desalinization in the Leziria Grande Polder, Portugal. Thesis. Wageningen Agricultural University, The Netherlands
22. ILRI, 2000. Irrigation, groundwater, drainage and soil salinity control in the alluvial fan of Garmsar. Consultancy assignment to the Food and Agriculture Organization (FAO) of the United Nations, International Institute for Land Reclamation and Improvement (ILRI), Wageningen, The Netherlands. On line:
23. SWAP model
24. DrainMod-S model
25. UnSatChem model
26. Hydrus model
27. LeachMod

Index

P

R

S

T

U

W